Nitrosation

Nitrosation

D.L.H. WILLIAMS

Reader in Chemistry, Durham University

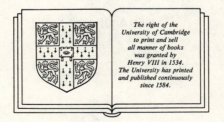

The right of the
University of Cambridge
to print and sell
all manner of books
was granted by
Henry VIII in 1534.
The University has printed
and published continuously
since 1584.

CAMBRIDGE UNIVERSITY PRESS

*Cambridge   New York   New Rochelle
Melbourne   Sydney*

CHEMISTRY

Published by the Press Syndicate of the University of Cambridge
The Pitt Building, Trumpington Street, Cambridge CB2 1RP
32 East 57th Street, New York, NY 10022, USA
10 Stamford Road, Oakleigh, Melbourne 3166, Australia

First published 1988

Printed in Great Britain at the University Press, Cambridge

British Library Cataloguing in Publication Data
Williams, D.L.H.
Nitrosation.
1. Nitrosation
I. Title
547'.26   QD281.N5

Library of Congress Cataloguing in Publication Data
Williams, D.L.H. (Daniel Lyn Howell)
Nitrosation.
Bibliography:
Includes index.
1. Nitrosation.   I. Title.
QD281.N5W55   1987   547'.26   86-33399

ISBN 0 521 26796 X

TM

CONTENTS

INTRODUCTION

Nitrosation reactions have a long and varied history within organic chemistry. Piria in 1846 nitrosated aliphatic primary amines and isolated the deamination products; later in 1850 Hofmann carried out similar reactions with aromatic amines. The chemistry of diazonium ions began with the work of Griess in 1858, nitrosamines were characterised shortly afterwards and in 1886 Fischer and Hepp effected the rearrangement of aromatic nitrosamines. Nitrosation at carbon originated with the work of Victor Meyer in 1873 when nitrolic acids were produced from aliphatic nitro compounds, and in 1877 Tilden isolated nitroso chlorides from the reaction of nitrosyl chloride with alkenes. Since these early days there have of course been very many discoveries and developments in nitrosation chemistry. Many nitrosation reactions are now standard procedures, both in the laboratory and on the industrial scale, providing many important routes in synthesis. In some cases the nitrosation products are themselves useful directly, whereas in others they are intermediates in the synthesis of other products. Some nitrosations which are industrially important include the formation of azo dyes by diazotisation of aromatic and heterocyclic amines (and coupling), the formation of hydroxylamine by nitrosation of bisulphite ion (Raschig process), the production of ε-caprolactam (and hence nylon 6) by the nitrosation of cyclohexane derivatives, the use of nitroso compounds in rubber production, as well as the use of alkyl nitrites and metal nitrosyl complexes as vasodilators in medicine. Since the discovery in 1956 that nitrosamines are powerful carcinogens, there has been a very large increase in studies of nitrosamine chemistry, particularly their formation and reactions. This area continues to be one of major concern, since nitrosamines can readily be formed under acid conditions (eg in the human stomach) from many naturally occurring amines and ingested nitrate and nitrite. Much work is being directed at the mechanism of carcinogenesis brought about by nitrosamines. Some nitrosamides have useful chemotherapy properties.

Nitrosation chemistry (particularly *N*-nitrosation) has also been a fruitful area for the mechanistic organic chemist. Many reactions are now well understood, and in several cases the nature of the effective nitrosating species has been established, often by kinetic procedures. Features such as acid catalysis, base catalysis, nucleophilic catalysis, diffusion-controlled processes, rate-limiting proton transfers and intramolecular rearrangements, have all been established in nitrosation reactions under various experimental conditions.

However, nitrosation has tended to be overshadowed somewhat by the possibly more well-known nitration reactions, and although there have been a number of important review articles and book chapters on certain areas of nitrosation chemistry, there has been no single book devoted to the subject. This book is an attempt to fill this gap. An effort has been made to combine both the synthetic and mechanistic aspects of nitrosation. The emphasis throughout is on the chemistry of the formation of nitroso compounds. Reactions of nitroso compounds are also in many cases very important, but are not discussed here, other than the cases where nitroso compounds such as alkyl nitrites, nitrosamines and some metal nitrosyl complexes act themselves as nitrosating agents ie where they transfer the nitroso group to another reactant.

The opening chapter discusses the range of reagents available which can bring about nitrosation, and subsequent chapters deal respectively with reactions which occur at aliphatic carbon, aromatic carbon, nitrogen, oxygen and sulphur sites within molecules, as well as those which involve metal complexes. There is an extremely large literature in the area of nitrosation chemistry; this book attempts to discuss reactions which are representative of their class, rather than to be comprehensive. Reaction mechanisms are given in cases where they have been properly established.

I would very much like to thank Professor Ken Schofield, not only for his valued editorial help with this book, but also for his encouragement to embark on this project. Thanks are also due to Professor John Ridd, Dr Geoffrey Stedman, and Dr Martin Hughes, who have read various chapters, and who have made a number of valuable suggestions. Errors and omissions remain my responsibility.

Durham, September 1986   D.L.H. WILLIAMS

1

Reagents involved in nitrosation

1.1 Acidic solutions of nitrous acid

By far the most widely used reagent in nitrosation and diazotisation is that derived from nitrous acid, solutions of which are readily produced from sodium nitrite (or any other nitrite salt) and aqueous mineral acid. Nitrous acid itself is only commonly known in solution (but has been detected in the gas phase), and decomposes fairly readily in the presence of acid, as outlined in equation (1.1). Solutions are therefore used immediately, and quantitative work has to take account of the decomposition. However, many nitrosation reactions are sufficiently rapid at 25 °C or 0 °C (where most of the kinetic work has been carried out), to allow the decomposition reaction to be ignored.

$$3HNO_2 = 2NO + HNO_3 + H_2O \qquad (1.1)$$

The structure of the nitrous acid molecule is well established.[1] It exists in *cis* and *trans* forms, with the *trans* form dominant. The bond lengths are, N=O 1.20 Å, O—N 1.46 Å, and O—H 0.98 Å, and the bond angles are as shown. The dipolar character of the molecule, as depicted in equation (1.2), contributes to the overall structure, and accounts for the partial double-bond nature of the HO—N bond indicated by the bond length.

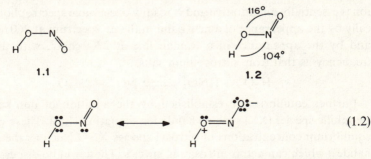

$$(1.2)$$

Nitrous acid is a weak acid, its pK_a was first established conductometrically[2] as 3.35 at 25 °C. Later, many workers have reported similar values derived from potentiometric, conductometric, spectrophotometric and kinetic measurements. Probably the most reliable values are those of Lumme and Tummavuori[3] who examined the variation with ionic strength, and quote a thermodynamic value of 3.148 at zero ionic strength at 25 °C. Thus in any quantitative work at pH $< \sim 2$ the ionisation to nitrite ion can be ignored.

$$2HNO_2 \rightleftharpoons N_2O_3 + H_2O \qquad (1.3)$$

Another important equilibrium exists in aqueous solution, involving the formation of dinitrogen trioxide N_2O_3 as shown in equation (1.3). At reasonably high nitrous acid concentrations (*ca* 0.1 M), and at moderate acidities (4 M), substantial quantities of dinitrogen trioxide (absorption maximum 625 nm) are formed, and the blue colour is easily detected by eye. The value of the equilibrium constant for its formation has, however, been the subject of some uncertainty. Early work using both spectrophotometric and distribution methods[4] gave a value of *ca* 0.2 l mol^{-1}, but more recent experiments by Markovits *et al.*,[5] claim that the correct value is 3.0×10^{-3} l mol^{-1}. The large discrepancy is not easy to explain, but may result from the quite high acid concentrations necessarily used in the early experiments. The more recent value agrees rather better with one calculated from thermodynamic data, and is based on a measured extinction coefficient in the uv for dinitrogen trioxide which agrees with a value determined by pulse radiolysis. The importance of a reliable value for the equilibrium constant will be evident later in this section, when bimolecular rate constants for dinitrogen trioxide reactions are discussed.

At much higher acidities nitrous acid is converted into the nitrosonium ion NO^+ (equation (1.4))[6]. Nitrosonium salts are now well characterised and their utility as nitrosating agents will be discussed later in section 1.5. In 60% perchloric acid or 60% sulphuric acid the conversion to nitrosonium ion is essentially quantitative and is readily observable spectrophotometrically by the appearance of a new band in the uv spectrum at 260 nm,[6(a),(b)] and by the appearance of a Raman line at 2300 cm^{-1}, with the same frequency as that from nitrosonium salts.[7]

$$H_3O^+ + HNO_2 \rightleftharpoons NO^+ + 2H_2O \qquad (1.4)$$

Further equilibria are established by the addition of non-basic nucleophilic species (X^-) such as halide ion (equation (1.5)). These generate equilibrium concentrations of nitrosyl species XNO (such as the nitrosyl halides) which can act as nitrosating species. These will be discussed fully later in section 1.2.

$$HNO_2 + H_3O^+ + X^- \rightleftharpoons XNO + 2H_2O \qquad (1.5)$$

The remainder of this section will be concerned with nitrosation reactions of aqueous acidic solutions of nitrous acid, in the absence of any added nucleophilic species. This falls into two parts: reaction via dinitrogen trioxide generated in solution, and acid-catalysed pathways.

1.1.1 Reaction via dinitrogen trioxide

Dinitrogen trioxide in solution or in the gas phase can effect nitrosation. This latter aspect will be discussed in section 1.7. Here we are concerned with reactions whereby dinitrogen trioxide is the reactive species generated in acidified aqueous solutions of inorganic nitrite. By far the most studied reactions in this category involve reactions of amines. The early literature is confused, but can now readily be rationalised. An excellent coverage of the early work together with mechanistic explanation was given in a review by Ridd in 1961.[8] Reaction involves attack of a substrate S by dinitrogen trioxide, generated from nitrous acid as outlined in scheme (1.6). The expected rate equation is thus equation (1.7) which predicts a second-order dependence upon the nitrous acid concentration and a first-order dependence upon the substrate concentration. This rate equation has been observed for a wide range of substrates. Historically, the mechanistic interpretation of the experimental results was given by Hammett.[9] Another explanation of the third-order rate equation by Kenner[10] involved rapid reversible attack by nitrous acidium ion $H_2NO_2^+$ and a subsequent rate-limiting proton transfer to the nitrite ion, which was later shown to be incorrect on a number of counts.

$$\left. \begin{aligned} 2HNO_2 &\stackrel{K}{\rightleftharpoons} N_2O_3 + H_2O \\ N_2O_3 + S &\stackrel{k}{\longrightarrow} \overset{+}{S}\!-\!NO + NO_2^- \end{aligned} \right\} \tag{1.6}$$

$$\text{Rate} = k[N_2O_3][S] = kK[HNO_2]^2[S] \tag{1.7}$$

$$\left. \begin{aligned} 2HNO_2 &\stackrel{k'}{\longrightarrow} N_2O_3 + H_2O \\ \text{Rate} &= k'[HNO_2]^2 \end{aligned} \right\} \tag{1.8}$$

For very reactive substrates (or substrates at high concentration) the reaction with dinitrogen trioxide can be faster than the hydrolysis of dinitrogen trioxide to nitrous acid, and so the rate-limiting step becomes the formation of dinitrogen trioxide (see equation (1.8)), the rate equation now being second-order in nitrous acid concentration and zero-order in substrate concentration. This limiting condition has been achieved using aniline,[11] N-methylaniline,[12] 1, 2-dimethylindole,[13] ascorbic acid,[14] azide

ion,[15] hydroxylamine,[16] piperazine[17] and thiosulphate ion.[18] The values of
k' (equation (1.8)) vary somewhat with the different conditions employed
but are approximately constant at $ca\,9\,l\,mol^{-1}\,s^{-1}$ at $25\,^\circ C$ and
$0.8\,l\,mol^{-1}\,s^{-1}$ at $0\,^\circ C$. Confirmation of the mechanism comes from the
results of experiments involving ^{18}O exchange with the solvent.[8,19] In the
absence of a substrate ^{18}O exchange occurs between nitrous acid and the
water solvent. This exchange is second-order with respect to the nitrous
acid concentration and k' (equation (1.8)) is in good agreement with that
determined for diazotisation of aniline.

Further details of the mechanism of dinitrogen trioxide formation emerge
from the observation of base catalysis (when equation (1.8) applies). This has
been interpreted[8] in terms of the sequence set out in scheme (1.9) where the
protonated form of nitrous acid reacts rapidly with the buffer anion B^-,
giving BNO which reacts with nitrite ion in a rate-limiting step to give
dinitrogen trioxide. A similar scheme can be written involving nitrosonium
ion rather than nitrous acidium ion.

$$\left.\begin{aligned}
HNO_2 + H_3O^+ &\xrightleftharpoons{\text{fast}} H_2NO_2^+ + H_2O \\[2mm]
H_2NO_2^+ + B^- &\xrightleftharpoons{\text{fast}} BNO + H_2O \\[2mm]
BNO + NO_2^- &\xrightarrow{\text{slow}} N_2O_3 + B^-
\end{aligned}\right\} \qquad (1.9)$$

With less reactive substrates or for amines at higher acidity where the
concentration of the reactive free base is much reduced, equation (1.7) is
commonly encountered. So if the equilibrium constant for dinitrogen
trioxide formation (K) is known, then it is easy to deduce the rate constant k
for the dinitrogen trioxide reaction. Obviously the values of k so obtained
are crucially dependent on the choice for the value of K. Using the earlier
value of $0.2\,l\,mol^{-1}$, it had been noted that the deduced k values were very
insensitive to the reactivity of the amine as measured by its pK_a value.
Indeed for $pK_a > 5$, values of k are virtually constant. This would imply that
reaction occurs at the encounter-controlled limit, but paradoxically the
limiting k value is several powers of ten below that expected for such a
process. This problem disappears if one accepts the more recent value of
$3.0 \times 10^{-3}\,l\,mol^{-1}$ for dinitrogen trioxide formation. Values of k obtained
in this way are gathered together in table 1.1.

It is clear that N_2O_3 is a more reactive species than thought hitherto;[24]
the small range of rate constant values for amines with $pK_a > \sim 5$ is now
much closer to the calculated limit for encounter controlled reactions of

Table 1.1. *Values of k in Rate = k[N_2O_3][S] in water at 25°C (except where stated)*

Substrate	pK_a	k/l mol^{-1} s^{-1}	Reference
4-Trimethylammoniumaniline (0°C)	2.51	4.7×10^6	20
2-Chloroaniline	2.63	1.4×10^7	20
3-Chloroaniline	3.46	9.6×10^7	20
4-Chloroaniline	3.92	2.8×10^8	20
2-Methylaniline	4.39	4.2×10^8	20
Aniline	4.60	7.5×10^8	20
3-Methylaniline	4.69	8.2×10^8	20
N-Methylaniline	4.85	4.0×10^8	20
4-Methylaniline	5.07	1.9×10^9	20
4-Methoxyaniline (0°C)	5.29	1.8×10^8	20
Piperazine	5.55	1.3×10^8	20
Hydroxylamine	5.90	2.0×10^8	20
Mononitrosopiperazine	6.80	7.5×10^7	20
Morpholine	8.70	2.2×10^8	20
Methylbenzylamine	9.54	1.8×10^8	20
Ethylbenzylamine	9.68	3.1×10^8	20
N-Methylglycine	10.20	1.5×10^8	20
2-Butylamine	10.56	1.6×10^8	20
Propylamine	10.67	9.4×10^7	20
Methylamine	10.70	1.6×10^8	20
Diisobutylamine	10.82	1.3×10^8	20
Dimethylamine	10.87	1.2×10^8	20
Diethylamine	10.98	1.1×10^8	20
Dipropylamine	11.00	1.2×10^8	20
Methylcyclohexylamine	11.04	2.2×10^8	20
Diisopropylamine	11.20	1.6×10^8	20
Piperidine	11.20	1.3×10^8	20
Dibutylamine	11.25	1.8×10^8	20
2-Methylbut-2-ene (0°C)	—	5.5×10^4	21
2, 3-Dimethylbut-2-ene (0°C)	—	3.9×10^5	21
2-Phenylindole (3°C)	—	0.6×10^8	13
2-Phenylindolizine	—	9.3×10^9	22
Azide ion	—	2.1×10^9	23
3-Chlorophenylhydrazine	—	5.9×10^9	23

$\sim 7 \times 10^9 \, \text{l} \, \text{mol}^{-1} \, \text{s}^{-1}$ at 25 °C. Reactions with two other powerful nucleophiles, azide ion and 3-chlorophenylhydrazine also occur at the encounter limit. *C*-Nitrosation also can be effected by N_2O_3; the two alkenes 2-methylbut-2-ene and 2, 3-dimethylbut-2-ene react in the expected rate sequence for an electrophilic process. The very much more reactive systems, the indole and indolizine derivatives again react upon encounter. Confirmation that many of the amine reactions occur at the encounter limit emerges from the measurement of the activation energies for these reactions.[20] The values lie in the range $10-20 \, \text{kJ} \, \text{mol}^{-1}$, which is typical for reactions of this type.[25] Similar conclusions can be drawn for reactions of basic amines using gaseous N_2O_3 dissolved in aqueous alkali,[26] rather than N_2O_3 generated *in situ* from nitrous acid. Reactions of the nitrogen oxides in the gas phase are discussed in section 1.7 of this chapter.

1.1.2 Acid-catalysed pathways

At higher acidities and lower total nitrous acid concentration (when the concentration of free nitrite ion is low) other mechanisms operate, where the reagent is not now dinitrogen trioxide. This is easily detected experimentally because a different rate equation (equation (1.10)) comes into play, where reaction is now first-order in each of the nitrous acid, substrate and free hydrogen ion concentrations.

$$\text{Rate} = k[\text{HNO}_2][\text{H}_3\text{O}^+][\text{S}] \qquad (1.10)$$

This equation is very commonly encountered for a large range of structurally different substrates, including *N*-, *O*-, *C*-, and *S*-nitrosation of organic and inorganic species and also for the nitrosation of a number of anions. This mechanism was first reported[27] for the diazotisation of 2-chloroaniline in dilute perchloric acid, in one of a series of important papers in 1958 by Hughes, Ingold & Ridd, which established many mechanistic features of nitrosation. The mechanistic interpretation, however, has been the subject of some controversy over a number of years since two plausible mechanisms are consistent with rate equation (equation (1.10)). The first, proposed by Hughes, Ingold & Ridd[28] (given in scheme (1.11)) involves rate-limiting attack by the nitrous acidium ion (which can be regarded as the hydrated nitrosonium ion); the second has the nitrosonium ion itself as the reactive species, as outlined in scheme (1.12). Arguments for and against each mechanism will be presented, but it has not yet been unambiguously demonstrated which is correct. At very high acidities ($\sim 60\%$ perchloric acid) there is no doubt that the nitrosonium ion is the reactive entity. Its presence in solution has been clearly demonstrated and the equilibrium constant for its formation (equation (1.4)) has been variously determined[29] in the range 3.0×10^{-5}–$7.8 \times 10^{-9} \, l \, mol^{-1}$. The rate equation for diazotisation at these very high acidities (given in equation (1.13)) where h_0 is the Hammett acidity function, has been interpreted[30] in terms of a rapid equilibrium reaction between nitrosonium ion and the protonated amine, followed by a slow proton transfer. This is discussed more fully in section 1.5 and also in chapter 4.

$$\left.\begin{array}{l} HNO_2 + H_3O^+ \rightleftharpoons H_2NO_2^+ + H_2O \\ H_2NO_2^+ + S \longrightarrow Product \end{array}\right\} \qquad (1.11)$$

$$\left.\begin{array}{l} HNO_2 + H_3O^+ \rightleftharpoons H_2NO_2^+ + H_2O \\ H_2NO_2^+ \rightleftharpoons NO^+ + H_2O \\ NO^+ + S \longrightarrow Product \end{array}\right\} \qquad (1.12)$$

$$Rate = k[ArNH_3^+][NO^+]h_0^{-2} \qquad (1.13)$$

Reaction via nitrosonium ion is a reasonable suggestion based on the analogy with nitration involving the nitronium ion NO_2^+. However, perhaps the most convincing argument for the involvement of nitronium ion in nitration is the observation of a rate equation which is zero-order in the substrate, which has been found for many very reactive aromatic substrates. This then represents the rate-limiting formation of the nitronium ion. Similar claims have been made in nitrosation at low acidities, particularly in the case of the nitrosation of hydrogen peroxide[31] and of thiosulphate ion[32]. However, in neither case is the evidence totally convincing since other explanations are also possible. For example in the hydrogen peroxide case, curvature is found for the plot of the first-order rate constant against $[H_2O_2]$ (when $[H_2O_2] \gg [HNO_2]$); the fully zero-order dependence is not achieved, and further, very high concentrations of substrate (approaching 1 M) are required to bring about this curvature. At these high concentrations the curvature could also result from a medium effect. This has been shown to be the case in the nitration of some reactive aromatic systems in acetic anhydride.[33] Similarly in the nitrosation of alcohols the observed curvature for the same plots is clearly a medium effect, which can also be brought about by the addition of tetrahydrofuran, which is a non-reacting material.[34] Again, for the thiosulphate reaction the claimed zero-order component to the rate equation could be due to rate-limiting dinitrogen trioxide formation.[18]

$$\text{Rate} = k[\text{HNO}_2][\text{H}^+] \tag{1.14}$$

One of the main arguments against the nitrosonium ion as the electrophile in dilute acid solution comes from the result of ^{18}O exchange experiments.[19] Exchange between nitrous acid and water occurs according to rate equation (1.14), with a value for k of 230 $\text{l mol}^{-1}\text{s}^{-1}$ at 0 °C. If the nitrosonium ion is involved (as in scheme (1.12)) then the exchange of one oxygen atom in nitrous acid occurs for each nitrosonium ion formed which then undergoes rehydration. This must necessarily be faster than the subsequent reaction with a substrate S if rate equation (1.10) is obeyed, yet for a range of anions, including azide ion, there is a limiting value for k of *ca* 2500 $\text{l mol}^{-1}\text{s}^{-1}$ at 0 °C. So at $[\text{N}_3^-] > \sim 0.1\,\text{M}$ the final step $\text{NO}^+ + \text{N}_3^- \rightarrow$ product is faster than is the formation of nitrosonium ion, which is at odds with the observed rate equation (equation (1.10)), where there is a first-order azide ion dependence.

The original mechanistic interpretation[28] of equation (1.10) was given in terms of scheme (1.11) where the reactive species is the nitrous acidium ion. This species had been proposed earlier[6(b)] as one present in acidified nitrous acid solution, even though it has not been detected spectro-photometrically. It does nevertheless represent a reasonable possibility as the effective reagent is dilute aqueous acid solution. There still remains the possibility though, that the nitrosonium ion is reactive under these conditions; if it has a very short lifetime (half-life $\sim 3 \times 10^{-10}$ s) then on rehydration it could well react with the same water molecule to which it was previously bound, thus not allowing ^{18}O exchange.

At the present time there appears to be no absolute clear-cut experimental evidence which differentiates between NO^+ and H_2NO_2^+ as the reactive species. Another approach[35] has been to envisage the structures **1.3** and **1.4** as representing limiting forms of the true structure of the electrophile. The difference between them being that in **1.3** one discrete water molecule is covalently bound to the nitroso group, whereas in **1.4** it is bound by solvation. The contribution of each form would be expected to vary with the acidity of the medium.

$$\text{H}-\overset{+}{\text{O}}-\text{N}=\text{O} \qquad\qquad \text{H}-\text{O}---\text{N}=\overset{+}{\text{O}}$$
$$\quad\;\;\vert \qquad\qquad\qquad\qquad\quad\;\;\vert$$
$$\quad\;\;\text{H} \qquad\qquad\qquad\qquad\quad\;\;\text{H}$$
$$\qquad\textbf{1.3} \qquad\qquad\qquad\qquad\quad\textbf{1.4}$$

The results of two independent theoretical studies concerned with nitrous acid protonation and nitrosonium ion formation have been published. In one case[36(a)] the most favourable conformation of the nitrous acidium ion is predicted by *ab initio* molecular orbital calculations, to be that where protonation occurs at the hydroxylic oxygen atom. This structure has an unusually long N—O(2) bond distance, and can be thought of as a complex between nitrosonium ion and water. The other treatment[36(b)] is based on the concept of charge- and frontier-orbital-controlled reactions, and concludes that in dilute acid solution nitrosation occurs via a nitrosonium ion–water complex, where the bonding is coulombic, rather than via the free nitrosonium cation.

Whatever the true nature of the electrophile, it is of interest to compare the reactivity of substrates towards nitrous acid. Some values of the third-order rate constant k (from equation (1.10)) are given in table 1.2. This is not meant as a comprehensive list, but examples have been selected from the literature to illustrate the range of reactivity. Further values are given in

Table 1.2. *Values of k in Rate $= k[HNO_2][H_3O^+][S]$*

Substrate	Temperature/°C	$k/l^2\,mol^{-2}\,s^{-1}$	References
Urea	25	0.89	37
Methylurea	25	10	38
Hydrazoic acid	25	160	37
Hydrazinium ion	25	620, 611	37, 39
Methanol	25	700	40
Sulphamate ion	25	1167	41
2, 4-Dinitroaniline	25	2.5	37
4-Nitroaniline	25	2 600	37
Aniline	25	4 600	42
4-Methylaniline	25	6 000	42
Cysteine	25	456, 443	43, 44
Mercaptopropanoic acid	25	4 764	44
Thiourea	25	6 960	43
Chloride ion	0	975	15
Bromide ion	0	1 170	45
Iodide ion	0	1 370	45
Thiocyanate ion	0	1 460	15
Nitrite ion	0	1 893	11
Ascorbate ion	0	2 000	14
Acetate ion	0	2 200	46
Azide ion	0	2 340	15
Thiocyanate ion	25	11 700	37
Benzenesulphinate ion	25	11 800	47
Thiosulphate ion	25	18 000	18

tables in later chapters when specific substrates are treated under the headings of N-, C-, O-, and S-nitrosation.

In general, the data in table 1.2 are consistent with an electrophilic process. One point of interest is that k values for reactive species tend to a limiting value. For neutral substrates this occurs at ca $7000 \, l^2 \, mol^{-2} \, s^{-1}$ at $25 \, ^\circ C$, which is taken to be the encounter-controlled limit. In the case of the anionic substrates, there is only a relatively small spread of values around $2000 \, l^2 \, mol^{-2} \, s^{-1}$ at $0 \, ^\circ C$ which again is taken as the encounter limit. This limit is, as expected, a little higher than for neutral substrates when compared at the same temperature ($11 \, 700$ against $7000 \, l^2 \, mol^{-2} \, s^{-1}$), due to the additional electrostatic factors. For a doubly-charged anion the limit is probably even greater at ca $18 \, 000 \, l^2 \, mol^{-2} \, s^{-1}$, but there is only one such example reported in the literature.

Most of the data for the anion reactions were determined indirectly, using a very reactive nitrostable substrate such as azide ion or aniline in the presence of the anion X^-. If the substrate concentration is sufficient to ensure that its rate of reaction is much greater than the hydrolysis of XNO (see scheme (1.15)), then XNO formation is rate-limiting. In quantitative terms this occurs when $k_2[S] \gg k_{-1}[H_2O]$ in scheme (1.15).

$$
\left.
\begin{aligned}
HNO_2 + H_3O^+ &\rightleftharpoons H_2NO_2^+ + H_2O \\
H_2NO_2^+ + X^- &\underset{k_{-1}}{\overset{k_1}{\rightleftharpoons}} XNO + H_2O \\
XNO + S &\xrightarrow{k_2} Product
\end{aligned}
\right\} \qquad (1.15)
$$

It is possible that an encounter pair $S.H_2NO_2^+$ (for a general substrate S) might be formed by a pre-association mechanism, rather than one involving a direct encounter-controlled reaction between S and nitrous acidium ion. Pre-association of S and nitrous acid would allow an alternative pathway for the formation of $S.H_2NO_2^+$. Detailed arguments in favour of this mechanism have been presented by Ridd (reference 25 pp. 41–5) for the bromination of aromatic substrates via the protonated form of hypobromous acid $BrOH_2^+$. There are many similarities between acid-catalysed nitrosation reactions and acid-catalysed bromination reactions, and a pre-association mechanism would remove some of the difficulties associated with the former, discussed earlier in this section.

1.2 Nitrosyl halides

Nitrosyl halides, particularly nitrosyl chloride have also been much used as electrophilic nitrosating agents. Reaction can be conveniently carried out by dissolving the gas in any of a range of organic solvents such as ether, acetic acid, chloroform, carbon tetrachloride, toluene, nitrobenzene etc.[48] More recently it has been shown that aqueous solutions of nitrosyl chloride under neutral or alkaline conditions can also be effective.[49] The procedure in organic solvents is particularly useful for the reaction of substrates with a low solubility in water, which makes the nitrous acid procedure inappropriate. The nitrosyl halides have also been identified kinetically as the reactive species (generated *in situ*) in nitrosation reactions using nitrous acid in the presence of halide ion.

$$2NO + X_2 = 2XNO \qquad (1.16)$$

Nitrosyl fluoride, nitrosyl chloride and nitrosyl bromide are all well-characterised compounds, and nitrosyl iodide has also been detected in solution and in the gas phase spectroscopically and kinetically. Their preparation usually involves a direct reaction of nitric oxide with the halogen (equation 1.16). Some of their structural and physical properties[50] are given in table 1.3. All are non-linear and gaseous at room temperature and pressure. Decomposition to nitric oxide and the halogen occurs fairly readily and extensively, and hydrolysis (equation (1.17)) is also rapid. Liquid nitrosyl chloride has some use as an ionising solvent, in for example the formation of some nitrosonium salts (see equation (1.18)).

$$XNO + H_2O = HNO_2 + H^+ + X^- \qquad (1.17)$$

$$ClNO + FeCl_3 \xrightarrow{\text{liquid ClNO}} NO^+FeCl_4^- \qquad (1.18)$$

Nitrosyl chloride has been widely used to nitrosate a range of substrates. Examples are shown in equations (1.19) – (1.22) of reactions with alkenes, carbonyl compounds, primary amines and alcohols.

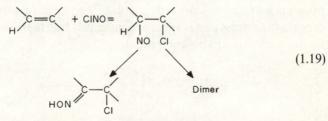

$$(1.19)$$

$$Me_2CO + ClNO = MeCOCH{=}NOH + HCl \qquad (1.20)$$

$$RNH_2 + ClNO \longrightarrow RN_2^+ \longrightarrow RCl \qquad (1.21)$$

$$ROH + ClNO = RONO + HCl \qquad (1.22)$$

The reaction with alkenes has been much used synthetically, particularly in terpene chemistry to characterise alkenes by identification of the nitroso chloride product, which generally exists as the oxime or the dimer.

Table 1.3. *Some structural and physical properties of the nitrosyl halides*

| Compound | Bond lengths/Å | | Bond Angles/° | Boiling Point/°C |
	N—X	N—O		
Nitrosyl fluoride	1.52	1.13	110	−59.9
Nitrosyl chloride	1.98	1.14	113	−6.4
Nitrosyl bromide	2.14	1.15	117	∼0

The formation of alkyl halides from primary aliphatic amines (equation (1.21)) is often accompanied by a number of other side reactions of the intermediate diazonium ion; this can be eliminated by generating the nitrosyl chloride *in situ* from an alkyl nitrite and titanium tetrachloride (equation (1.23)) in dimethylformamide at 0 °C.[51] Another convenient way to generate nitrosyl halides *in situ* in chlorinated hydrocarbon solvents is to treat an alkyl nitrite with the trimethylsilyl halide[52] (equation (1.24)).

$$TiCl_4 + 4RONO = Ti(OR)_4 + 4ClNO \qquad (1.23)$$
$$(CH_3)_3SiX + RONO = XNO + (CH_3)_3SiOR$$
$$(X = Cl, Br, I) \qquad (1.24)$$

C-Nitroso compounds can also be prepared from nitrosyl chloride under free-radical conditions by uv-irradiation, as exemplified in equation (1.25). The mechanism has not been established in detail, but it is reasonable to assume that radicals are involved. Thus from cyclohexane the nitroso dimer is sometimes formed, whereas in the presence of acid, the product is the oxime which can rearrange to give the caprolactam.[53] This is the basis of one industrial process for the synthesis of caprolactam which on ring-opening polymerisation yields nylon 6. As expected from a free-radical process the nitrosation is often accompanied by other reactions including the formation of chloro compounds.

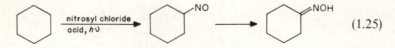

$$(1.25)$$

Probably the most commonly encountered reactions involving nitrosyl halides as nitrosating agents occur when these reagents are generated *in situ*, usually in aqueous acidic nitrous acid solutions containing halide ion. Under these conditions equilibrium concentrations of the nitrosyl halides are formed which then act as the reagent. The equilibrium constants for both nitrosyl chloride and nitrosyl bromide formation (defined by equ-

Table 1.4. *Equilibrium constants* $(K_{XNO}$ *in* $l^2\, mol^{-2})$ *for nitrosyl chloride and nitrosyl bromide formation in water*

	$K_{XNO}/0\,°C$	$K_{XNO}/25\,°C$
Nitrosyl chloride	5.6×10^{-4}	1.1×10^{-3}
Nitrosyl bromide	2.2×10^{-2}	5.1×10^{-2}

ations (1.26)) have been determined spectrophotometrically by Schmid and coworkers,[54] and are given in table 1.4. As expected from the greater nucleophilicity of the bromide ion, the K_{XNO} values are larger for nitrosyl bromide than for nitrosyl chloride. Values for nitrosyl fluoride and nitrosyl iodide have not been determined; the latter is too unstable with respect to iodine formation. In other solvents, K_{XNO} values are generally larger, for example in methanol[55] (see equation (1.27)) the values are 5×10^{-2} and 2.0 $l^2\, mol^{-2}$ for nitrosyl chloride and nitrosyl bromide respectively at $0\,°C$.

$$\left.\begin{array}{c} HNO_2 + X^- + H^+ \rightleftharpoons XNO + H_2O \\ K_{XNO} = \dfrac{[XNO]}{[HNO_2][X^-][H^+]} \end{array}\right\} \quad (1.26)$$

$$RONO + X^- + H^+ \rightleftharpoons XNO + ROH \quad (1.27)$$

The intervention of nitrosyl halides in aqueous (and other solvent) nitrosations can readily be deduced from the catalytic effects of added halide ion. Catalysis (eg of diazotisation) by hydrochloric acid has been long known and the rate equation (equation (1.28)), (where $[ArNH_2]$ is [unprotonated amine]), was first identified by Schmid in 1937.[56] This was later interpreted mechanistically by Hammett[9] in terms of rate-limiting attack on the free (unprotonated) amine by the nitrosyl halide, as outlined in scheme (1.29). The rate equation expected from such a scheme is given in equation (1.30), where K_A is the acid dissociation constant of $ArNH_3^+$ and $[HNO_2]_T$ and $[ArNH_2]_T$ are the total stoichiometric concentrations of the nitrous acid and the amine respectively. This assumes that $[ArNH_3^+] \approx [ArNH_2]_T$ and that $[XNO] \ll [HNO_2]_T$ and that the ionisation of the nitrous acid to nitrite can be neglected. Since K_{XNO} values are so small $[XNO]$ will always be $\ll [HNO_2]_T$ at all likely $[X^-]$ and $[H^+]$, but the first assumption holds only for the more basic amines. For less basic species, and indeed for non-basic reactants such as alcohols or thiols appropriate modification is necessary. The derived equation (equation

Table 1.5. *Rate constant values for the formation of the nitrosyl halides at $0\,^{\circ}C$*

X^-	$k/l^2\,mol^{-2}\,s^{-1}$	Reference
Cl^-	975	15
Br^-	1170	45
I^-	1370	45

(1.30)) is of the same form as that observed experimentally (equation (1.28)).

$$\text{Rate} = k[ArNH_2][H^+][HNO_2][X^-] \tag{1.28}$$

$$HNO_2 + X^- + H^+ \rightleftharpoons XNO + H_2O$$

$$XNO + ArNH_2 \xrightarrow[k_2]{slow} Ar\overset{+}{N}H_2NO \xrightarrow{fast} ArN_2^+ \tag{1.29}$$

$$\begin{aligned}\text{Rate} &= k[XNO][ArNH_2]\\ &= k_2K_{XNO}K_A[X^-][HNO_2]_T[ArNH_2]_T\end{aligned} \tag{1.30}$$

For some very reactive substrates the rate-limiting stage is changed to that of nitrosyl halide formation, just as for the dinitrogen trioxide reactions. This has been achieved with aniline derivatives[45] and azide ion[15] at low acidities, and with some sulphur compounds.[57] Mechanistically this can be envisaged[11] as rate-limiting attack by the halide ion on the nitrous acidium ion, or with the free nitrosonium ion. Either way, the third-order rate constants k in equation (1.31) can readily be obtained, and are given in table 1.5 for the three halide ions. Not only are the values very similar, given the large nucleophilicity differences, but they are also very close to values for a range of other anions,[4] again suggesting strongly that these reactions occur at the encounter limit.

$$\text{Rate} = k[H^+][HNO_2][X^-] \tag{1.31}$$

When nitrosyl halide attack is rate-limiting, it is a relatively easy matter to obtain the bimolecular rate constant k_2 (equation (1.30)) provided K_{XNO} and K_A are accurately known. Some of the results from the literature are shown in table 1.6. There has been no reported fluoride ion catalysis, and although iodide ion catalysis is quite marked, K_{INO} is not known, so the data in table 1.6 are limited to nitrosyl chloride and nitrosyl bromide reactions.

Nitrosyl chloride is always more reactive than nitrosyl bromide, as expected for an electrophilic nitrosation on the grounds of the electronegat-

Table 1.6. *Rate constants* $k_2 (1\,mol^{-1}s^{-1})$ *for the reactions of nitrosyl chloride and nitrosyl bromide in water at 25 °C (except where stated)*

Reactant	k_2(ClNO)	Reference	k_2(BrNO)	Reference
Methanol	2.1×10^5	40	2.0×10^4	40
Cysteine	1.2×10^6	58	5.8×10^4	58
Thioglycolic acid	1.4×10^7	58	1.1×10^6	58
Benzenesulphinic acid	4.6×10^7	58	9.5×10^6	58
Benzenesulphinate ion	2.4×10^8	58	1.2×10^7	58
Dimethylamine	3.1×10^7	59	3.6×10^7	59
Glycine	1.7×10^7	60	—	
4-Nitroaniline	2.1×10^8	61	4.3×10^7	61
4-Chloroaniline	1.9×10^9	62	—	
	1.8×10^9	61	2.5×10^9	61
Sulphanilic acid	1.4×10^9	37	9.9×10^8	37
Aniline	2.6×10^9	62	3.2×10^9	62
	2.5×10^9	61	1.7×10^9	61
4-Methylaniline	3.0×10^9	62	—	
	3.4×10^9	61	2.5×10^9	61
4-Methoxyaniline	5.1×10^9	61	2.8×10^9	61
Hydroxylamine (0 °C)	3.5×10^7	63	3.7×10^7	63
1, 2-Dimethylindole				
(3 °C)	9.9×10^8	64	—	
Azide ion (0 °C)	1.8×10^6	15	—	
Morpholine (31 °C)	—		4.7×10^7	65
N-Methylaniline	—		5.0×10^9	66

ivity difference between chlorine and bromine. Bromide ion catalysis, however, is always greater than chloride ion catalysis, because the difference in the K_{XNO} values outweighs the difference in the k_2 values. The variation of k_2 with substrate structure is again in line with an electrophilic process. For the more reactive species k_2 approaches the encounter-controlled limit $(7.4 \times 10^9\ 1\,mol^{-1}s^{-1})$ for both nitrosyl chloride and nitrosyl bromide eg for 4-methoxyaniline. This point will be further discussed in chapter 4.

$$S + XNO \underset{k_{-2}}{\overset{k_2}{\rightleftharpoons}} \overset{+}{S}{-}NO + X^-$$
$$\downarrow k_3$$

(1.32)

Final product

In some cases there is evidence that the initial nitrosation reaction is reversible, competing with the subsequent reactions of the nitroso ion first formed. In kinetic terms, reversibility will become important when $k_{-2}[X^-]$ is comparable with k_3. Experimentally this is detected by a curved plot for the observed rate constant *vs* the halide ion concentration $[X^-]$. This has been demonstrated particularly for the diazotisation of aniline derivatives.[55,61] As expected from scheme (1.32), reversibility is more important as the concentration of halide ion is increased, is more pronounced for the more powerful nucleophile ($Br^- > Cl^-$) and less basic substrates, and is more dominant in less polar solvents. A detailed kinetic analysis (including measurement and rationalisation of $k_{-2}:k_3$ ratios) is consistent with this interpretation. When the limit $k_{-2}[X^-] \gg k_3$ is achieved, by whatever means, then catalysis by halide ion should disappear. This is borne out experimentally for the much less basic (and less nucleophilic) substrates such as 2, 4-dinitroaniline and amides generally. Under these circumstances the further reaction of $\overset{+}{S}$—NO (often a proton transfer to the solvent) is rate-limiting. This explanation answers the question as to why amides (in particular) are not subjected to halide ion catalysis in nitrosation reactions, whereas for most amines the catalysis is generally very marked.

$$RONO + H^+ + X^- \rightleftharpoons ROH + XNO$$
$$XNO + \;{\Large\diagdown}NH \rightleftharpoons \overset{+}{N}—NO \longrightarrow Product \tag{1.33}$$

The *in situ* generation and subsequent reaction of nitrosyl halide also occur in solvents other than water. For example alkyl nitrites in acidic alcohol solvents are only effective nitrosating agents towards amines in the presence of halide ion (or other nucleophiles such as thiocyanate ion). Thus the diazotisation of aniline or the nitrosation of *N*-methylaniline by propyl nitrite in propanol, occurs at a measurable rate, only when nucleophilic catalysts are present.[67] The curved dependence of rate constant upon halide ion concentration is consistent with the reversibility of the initial nitrosation as outlined in scheme (1.33). Reversibility is more pronounced in the less polar solvent as expected.

1.3 Nitrosyl thiocyanate

Nitrosyl thiocyanate is not known as the pure compound, but exists as an unstable blood-red species in solution at low temperature. Its structure therefore has not been established beyond doubt because of its instability, but there are good reasons to believe that it exists as a covalent compound, in which the nitroso group is bound to the sulphur atom. This is to be expected from the principles of Hard-Acid-Soft-Base (HASB) theory, and a recent *ab initio* molecular orbital study[68] predicts that this structure is substantially more stable than the isomeric nitrosyl isothiocyanate.

The formation of nitrosyl thiocyanate solutions is discussed more fully in chapter 7 as an example of S-nitrosation. Such solutions prepared from nitrosyl chloride and silver thiocyanate or from ethyl nitrite and thiocyanic acid (see scheme (1.34)) have been used in a number of inorganic reactions,[69] but not specifically nitrosation reactions. Indeed such solutions in organic solvents do not appear to have any synthetic advantages over other nitrosating agents eg solutions of nitrosyl chloride.

$$\left.\begin{array}{l} ClNO + AgSCN = ONSCN + AgCl \\ RONO + HSCN = ONSCN + ROH \end{array}\right\} \tag{1.34}$$

However, like the nitrosyl halides (except for the fluoride), nitrosyl thiocyanate has been identified kinetically, as the effective nitrosating reagent for reactions carried out using aqueous acidic solutions of nitrous acid containing thiocyanate ion. Catalysis by thiocyanate ion was first noted for diazotisation of aniline[19] and for nitrosation of azide ion,[15] and has subsequently been observed for a large range of substrates. Reactions which are halide-ion-catalysed, are also catalysed by thiocyanate ion. In general, catalysis by thiocyanate is much more marked than is halide ion catalysis; for example, in the nitrosation of morpholine[65] the catalytic efficiency is $1:30:15\,000$, for chloride, bromide and thiocyanate ions respectively.

$$\left.\begin{array}{c} HNO_2 + SCN^- + H^+ \rightleftharpoons ONSCN + H_2O \\[2mm] K_{ONSCN} = \dfrac{[ONSCN]}{[HNO_2][H^+][SCN^-]} \\[2mm] ONSCN + S \xrightarrow{k} Product \end{array}\right\} \tag{1.35}$$

The origin of such a large effect lies in the magnitude of the equilibrium constant K_{ONSCN} for nitrosyl thiocyanate formation (scheme (1.35)). This has been determined[70] as $30\ l^2\ mol^{-2}$ at $25\,^\circ C$ and so is many orders of magnitude larger than the K_{ClNO} and K_{BrNO} values (table 1.4). The bimolecular rate constant (k in scheme (1.35)) for nitrosyl thiocyanate reaction is, in fact, smaller than for both nitrosyl halides, for all substrates which have been studied. Data for three such reactions, the diazotisation of aniline, the nitrosation of hydroxylamine, and the S-nitrosation of N-acetylpenicillamine, are shown in table 1.7.[34,63,71]

It is not immediately obvious why nitrosyl thiocyanate should be less

Table 1.7. *Rate constants* $(l\ mol^{-1}\ s^{-1})$ *for nitrosation in water by nitrosyl chloride, nitrosyl bromide and nitrosyl thiocyanate*

	Substrate		
	Aniline[a]	Hydroxylamine[b]	N-Acetyl-penicillamine[c]
Nitrosyl chloride	2.2×10^9	3.5×10^7	2.6×10^6
Nitrosyl bromide	1.7×10^9	3.7×10^7	1.4×10^5
Nitrosyl thiocyanate	1.9×10^8	3.6×10^6	3.0×10^3

[a]25 °C; [b]0 °C; [c]31 °C.

reactive than the nitrosyl halides, although there is the trend, the greater the nucleophilicity of the anion (X^-) then the lower is the reactivity of the nitrosyl compound (XNO). Molecular orbital calculations based on the frontier orbital approach[68] predict, however, that nitrosyl thiocyanate should be less reactive than nitrosyl chloride, which is borne out experimentally in every case studied. Again the large K_{ONSCN} value is responsible for the powerful catalytic effect of thiocyanate ion.

Thiocyanate catalysis of nitrosation of secondary amines may be an important feature of *in vivo* nitrosamine formation. Thiocyanate ions occur naturally in some vegetables and in human saliva (particularly in the saliva of smokers). Simulated *in vivo* experiments have shown[72] that, as expected, nitrosamine formation is substantially catalysed by typical naturally occurring concentrations of thiocyanate under gastric conditions (pH 1–2).

1.4 Nitrosyl acetate (acetyl nitrite)

A very common procedure for nitrosating substrates not soluble in water is to use sodium nitrite dissolved in glacial acetic acid. Under these circumstances it is believed that the effective nitrosating agent is nitrosyl acetate, formed according to equation (1.36). Nitrosyl acetate itself and other nitrosyl carboxylates are known compounds, their formation is discussed more fully in chapter 6 as examples of *O*-nitrosation.

$$CH_3COOH + HNO_2 = CH_3COONO + H_2O \qquad (1.36)$$

Again as for the halide and thiocyanate cases, catalysis of nitrosation by acetate and other carboxylate anions is observed; this is readily interpreted in terms of nitrosyl acetate formation. Nitrosyl acetate can react with the substrate, or with nitrite ion giving dinitrogen trioxide as the reactive species.[46] Since the nitrous acidium ion (or the nitrosonium ion) is also present there are now three possible nitrosating agents as shown in scheme (1.37). Each pathway has been identified kinetically[73] but the kinetic analysis is now necessarily more complex.

$$
\begin{array}{c}
HNO_2 + H^+ \rightleftharpoons H_2NO_2^+ \xrightarrow{\text{S}} \text{Product} \\
CH_3COO^- \nearrow \quad \nwarrow NO_2^- \\
CH_3COONO \underset{NO_2^-}{\rightleftharpoons} N_2O_3 \xrightarrow{\text{S}} \text{Product} \\
\quad (\text{S=Substrate}) \\
\downarrow \text{S} \\
\text{Product}
\end{array}
\tag{1.37}
$$

The equilibrium constant for nitrosyl acetate formation (equation (1.36)) cannot be measured directly as it is so small, but it has been estimated[73] to be *ca* 1.4×10^{-8} l mol^{-1}. This is deduced making use of the fact that nitrosation by nitrosyl acetate of nitrite ion, *N*-methylaniline, piperazine and azide ion (all relative to the rate constant for hydrolysis of nitrosyl acetate) all occur with about the same rate constant, implying that this is at the encounter limit.

1.5 Nitrosonium salts

Nitrosonium salts NO^+X^- are now well-known species, a number of which are commercially available. Some of the most commonly encountered are the tetrafluoroborate (BF_4^-), tetrachloroborate (BCl_4^-), hexafluorophosphate (PF_6^-), hydrogen sulphate (HSO_4^-), perchlorate (ClO_4^-) and the hexafluoroantimonate (SbF_6^-). Many others are known, some including metal complex anions eg $NO^+AlCl_4^-$ and $(NO^+)_2[Fe(CN)_5NO]^{2-}$. Under certain conditions dinitrogen trioxide and dinitrogen tetroxide react as if they were nitrosonium nitrite and nitrate respectively (and the nitrosonium cation can be detected spectroscopically), but are better treated as covalently bound species. In general nitrosonium salts are reasonably stable (although the perchlorate is explosive) crystalline materials, which are readily prepared from dinitrogen tetroxide, dinitrogen trioxide or nitrosyl chloride and a source of the anion, as outlined in equation (1.38) for nitrosonium hydrogen sulphate, which was also formed incidentally in the old lead chamber process for sulphuric acid manufacture.

$$
N_2O_4 + H_2SO_4 = NO^+HSO_4^- + HNO_3
\tag{1.38}
$$

As expected the salts are hydrolysed readily (to give nitrous acid) and so have to be used under anhydrous conditions. Their structures have been well established by X-ray crystallography,[74] vibrational spectroscopy[75] (NO stretching frequency 2250–2300 cm^{-1}) and by conductivity measurements.[76]

Predictably nitrosonium salts are very efficient nitrosating agents, and as such have been used quite widely synthetically. The procedure has been particularly useful for the isolation of diazonium ion salts[77] (generally from toluene solution) as outlined in equation (1.39).

$$NO^+BCl_4^- + ArNH_2 = ArN_2^+BCl_4^- + H_2O \tag{1.39}$$

Nitrosation of less reactive species (eg amides and sulphonamides) can readily be achieved with nitrosonium salts, where other reagents are not effective. Thus, primary aromatic amides are easily hydrolysed to the acids[78] (equation (1.40)) and secondary amides form the corresponding nitrosamides (equation (1.41)). Nitrosation of amides can also be achieved under basic conditions,[79] by N-metallation (with butyllithium) of the amide, followed by treatment with a nitrosonium salt in ether at 0 °C (equation (1.42)).

$$ArCONH_2 + NO^+BF_4^- = ArCO_2H + N_2 + HF + BF_3 \tag{1.40}$$

$$ArCON(R)H + NO^+BF_4^- = ArCON(NO)R + HF + BF_3 \tag{1.41}$$

$$RCON(R')H + BuLi \longrightarrow RCON(R')Li \xrightarrow{NO^+X^-} RCON(NO)R' \tag{1.42}$$

The reaction of amines with nitrosonium salts (in nitromethane or acetonitrile) in the presence of aromatic hydrocarbons leads to alkylation of the latter (equation 1.43)), presumably by way of initial diazotisation.[80]

$$RNH_2 + NO^+X^- \longrightarrow RN_2^+X^- \xrightarrow{ArH} ArR + N_2 + HX \tag{1.43}$$

The reactivity of nitrosonium salts (and other nitrosating agents) was much studied by Seel and coworkers[81] in a series of papers, 1950–7, including their reactions with azide ion in liquid sulphur dioxide (1.44) and with ethanol to give ethyl nitrite (1.45). Both are clear examples of nitrosation reactions.

$$NO^+ + N_3^- = N_2O + N_2 \tag{1.44}$$

$$NO^+ + C_2H_5OH = C_2H_5ONO + H^+ \tag{1.45}$$

Kinetic measurements of the diazotisation of aniline derivatives at high acidities indicate that the nitrosonium ion is the effective reagent at acidities above *ca* 6 M perchloric acid.[30] This is to be expected since in this acid region the bulk of the nitrous acid is present as the nitrosonium ion.[6] The rate equation (equation (1.46)) is very different from that observed at lower acidities. This, coupled with the large solvent isotope effect ($k_H/k_D \sim 10$)

suggests a sequence outlined in scheme (1.47), where rapid reversible nitrosation is followed by a slow proton transfer to the solvent S.

$$\text{Rate} = k[\text{ArNH}_3^+][\text{NO}^+]h_0^{-2} \tag{1.46}$$

$$
\left.
\begin{array}{l}
\text{ArNH}_3^+ + \text{NO}^+ \underset{\text{slow}}{\overset{\text{fast}}{\rightleftharpoons}} \text{Ar}\overset{+}{\text{N}}\text{H}_2\text{NO} + \text{H}^+ \\[2pt]
\text{Ar}\overset{+}{\text{N}}\text{H}_2\text{NO} + \text{S} \xrightarrow{\text{slow}} \text{ArNHNO} + \text{SH}^+ \\[4pt]
\qquad\qquad\qquad\Big\downarrow \text{fast} \\[6pt]
\qquad\qquad\qquad \text{ArN}_2^+
\end{array}
\right\} \tag{1.47}
$$

The nitrosonium ion is also the likely reactant in many diazotisation reactions carried out using nitrosonium hydrogen sulphate (or nitrosyl sulphuric acid) prepared by dissolving sodium nitrite in 90–96% sulphuric acid at 0 °C. This reagent is particularly useful for the diazotisation of weakly basic species such as polynitroanilines, aminoanthraquinones and many heterocyclic amines, many of which have been, or are currently, used industrially on a large scale in the synthesis of azo dyes.

It is worth noting that nitrosonium salts can, in addition to the two-electron transfer nitrosation reactions mentioned so far, effect one-electron transfer oxidation reactions. Two simple examples are given in equations (1.48) and (1.49) for the oxidation of iodide ion and sulphite ion, and are easily recognised by the formation of nitric oxide as one of the products. Reactions of this type have some synthetic utility in organic chemistry,[82] where nitrosonium salts are used as mild selective oxidising agents. Examples include selective oxidative cleavage of ethers and oximes, initiation of some electrophilic condensation reactions and also initiation of polymerisation reactions. Radical cations can readily be generated from aromatic substrates and nitrosonium salts, which can be isolated as their salts (see equations (1.50)).[83(a)] Another interesting example is the reaction of cyclic sulphides such as 1, 5-dithiocyclooctane with nitrosonium tetrafluoroborate in acetonitrile,[83(b)] which gives the long-lived radical cation (equation 1.51)), which can be isolated as the hexafluorophosphate salt; the radical shows evidence (from esr measurements) of sulphur–sulphur transannular interaction. An analogous N—S bond is formed when one sulphur atom of the cyclooctane is replaced by —NCH$_3$. The corresponding dications of both examples are formed if two equivalents of nitrosonium salt are allowed to react.

$$2\text{NO}^+ + 2\text{I}^- = 2\text{NO} + \text{I}_2 \tag{1.48}$$

$$2\text{NO}^+ + \text{SO}_3^{2-} = \text{SO}_3 + 2\text{NO} \tag{1.49}$$

$$Ar + NO^+BF_4^- \xrightarrow{\text{acetonitrile}} Ar^{\cdot+}BF_4^- + NO \qquad (1.50)$$

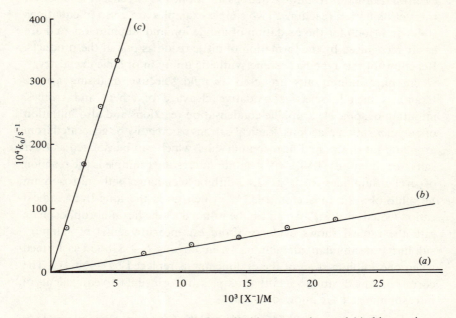

$$\qquad (1.51)$$

In some cases the one-electron transfer reaction occurs in preference to the expected two-electron transfer nitrosation reaction. Some diphenylamine derivatives, when treated with nitrosonium tetrafluoroborate in acetonitrile give, instead of the nitrosamine products, the radical cations which then dimerise (see equation (1.52)).[84]

$$Ar_2NH + NO^+ = Ar_2\overset{+\bullet}{N}H + NO$$

$$\qquad (1.52)$$

$$Ar_2NNAr_2 + 2H^+$$

1.6 Nitrosothiouronium ions

The formation of *S*-nitroso species derived from nitrous acid and thiourea derivatives is discussed in chapter 7 under *S*-nitrosation. The equilibrium

Figure 1.1 Catalysis by (*a*) bromide ion, (*b*) thiocyanate ion and (*c*) thiourea in the nitrosation of morpholine

Table 1.8. *Rate constants* $(l\,mol^{-1}\,s^{-1})$ *for diazotisation in water by nitrosyl chloride, nitrosyl bromide, nitrosyl thiocynate and S-nitrosothiouronium ion at 25 °C*

	Aniline	4-Aminobenzoic acid
Nitrosyl chloride	2.2×10^9	1.1×10^9
Nitrosyl bromide	1.7×10^9	4.3×10^8
Nitrosyl thiocyanate	1.9×10^8	1.4×10^6
S-Nitrosothiouronium ion	1.3×10^6	1.8×10^4

constant for the reaction of thiourea itself (equation (1.53)) is 5000 $l^2\,mol^{-2}$ at 25 °C.[85] The S-nitrosothiouronium ion is itself a nitrosating agent. This is clearly demonstrated by the catalysis of the nitrosation of morpholine by added thiourea.[86] Figure 1.1 shows the dependence of the first-order rate constant (with [morpholine] $\gg$ [nitrous acid]) upon the concentration of added thiourea. Data for the corresponding catalysis by thiocyanate ion and bromide ion are included for comparison purposes. Clearly thiourea catalysis is substantial, indeed it is probably the most efficient catalyst yet found for nitrosation. The results are readily explained, as for halide ion and thiocyanate ion catalysis, by the formation of an equilbrium concentration of the S-nitroso ion (equation (1.53)) which then directly nitrosates the secondary amine (equation (1.54)). Similar, substantial catalytic effects have been observed for the nitrosation of aliphatic amines,[87] as well as for the diazotisation of anilines;[34,86] the alkyl thioureas behave similarly. Kinetic analysis reveals that the actual reactivity of the S-nitrosothiouronium ion (as measured by its second-order rate constant for reaction with a substrate) is in fact much lower than that of nitrosyl chloride, nitrosyl bromide and nitrosyl thiocyanate, and the large catalytic effect arises from the very large equilibrium constant (K_{XNO}) for S-nitrosothiouronium ion formation. The trends are clearly seen for two substrates, aniline and 4-aminobenzoic acid in table 1.8.

$$H^+ + HNO_2 + (NH_2)_2CS \rightleftharpoons (NH_2)_2C\overset{+}{S}NO + H_2O \qquad (1.53)$$

$$(1.54)$$

Catalysis by thiourea seems to be a general feature for nitrosation reactions which also exhibit halide ion (and thiocyanate and acetate ion) catalysis, and has been demonstrated for the reactions of hydrazoic acid,[37] hydroxylamine,[37] and benzenesulphinic acid.[47]

1.7 Nitrogen oxides

On the preparative side a number of nitroso compounds have been synthesised using various oxides of nitrogen, often as a mixture of 'nitrous fumes' of rather ill-defined composition. Nitro compounds can also be obtained by a nitration process using some oxides of nitrogen; often the two processes occur together. Mechanistically the reactions are now reasonably well understood. The three relevant oxides so far as nitrosation is concerned are dinitrogen trioxide, dinitrogen tetroxide and nitric oxide; each will be considered in turn.

1.7.1 Dinitrogen trioxide N_2O_3

Dinitrogen trioxide (or nitrous anhydride) exists as a pure compound only as a blue solid or liquid at low temperatures. Dissociation into nitric oxide and nitrogen dioxide (equation (1.55)) occurs increasingly as the temperature rises, and some of the reactions of dinitrogen trioxide are in fact reactions of nitrogen dioxide. The preparation, properties and general reactions of dinitrogen trioxide are described in reference 88.

$$N_2O_3 \rightleftharpoons NO + NO_2 \qquad (1.55)$$

Reactions in water where dinitrogen trioxide exists in equilibrium with nitrous acid have already been discussed in section 1.1.1. This present section deals with reactions in the gas phase and also in solution in organic solvents. Dinitrogen trioxide dissolves readily in a large number of organic solvents, and these solutions have been used to nitrosate amines[89] (primary and secondary) and carbonyl compounds[90] as outlined in equations (1.56), (1.57) and (1.58). In addition alkyl nitrites and thionitrites have been prepared from alcohols and thiols in this way. The method does not, however, appear to have any major practical advantage over other procedures. The products of amine (and other substrate) nitrosations are essentially the same using dinitrogen trioxide as they are using nitrous acid, which suggests that the same mechanism of electrophilic nitrosation obtains, although there has been no major mechanistic study.

$$ArNH_2 + N_2O_3 \xrightarrow{\text{toluene}} ArN_2^+NO_2^- + H_2O \qquad (1.56)$$

$$R_2NH + N_2O_3 \xrightarrow{\text{toluene}} R_2NNO + HNO_2 \qquad (1.57)$$

$$C_6H_5COCH_2COC_6H_5 + N_2O_3 \xrightarrow{\text{ether}} C_6H_5COCHCOC_6H_5 + HNO_2$$
$$\underset{\displaystyle NO}{|} \qquad (1.58)$$

Solutions of secondary amines in acetonitrile rapidly yield nitrosamines when exposed to gaseous dinitrogen trioxide, and aqueous alkaline solutions of amines react similarly, and concurrently with hydrolysis of dinitrogen trioxide giving nitrite ion (see equation (1.59)[26,91]). The yield of nitrosamine (relative to that of nitrite ion) varies only little over a range of amines of very different basicities, implying that as for reactions in water, dinitrogen trioxide reacts very rapidly at the encounter limit. In these reactions no *N*-nitroamines are formed (contrasting with the reactions of dinitrogen tetroxide). This is taken as evidence against a radical mechanism involving attack by nitrogen dioxide, although some other reactions of dinitrogen trioxide do form nitro products which undoubtedly result from such a radical mechanism.

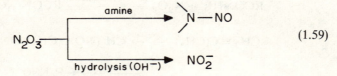

$$\qquad (1.59)$$

1.7.2 Dinitrogen tetroxide N_2O_4

The chemistry of dinitrogen tetroxide is closely bound up with the equilibrium reaction (equation (1.60)) forming nitrogen dioxide. Just as for dinitrogen trioxide a number of reactions are best rationalised in terms of a free-radical reaction generally involving attack by nitrogen dioxide and often resulting in nitro compounds.[92] However, many other reactions clearly involve nitrosation, probably via an electrophilic process. Details of preparation, structure, properties and general reactions of dinitrogen tetroxide are given in reference 93.

$$2NO_2 \rightleftharpoons N_2O_4 \qquad (1.60)$$

The structure of dinitrogen tetroxide, at least in the solid form is well established in terms of a planar molecule containing the N—N bond and a centre of symmetry.[93] However, many of its reactions are best interpreted[94] in terms of an ionic structure $NO^+NO_3^-$. There is excellent evidence for this ionic form, particularly in sulphuric and perchloric acids, where conversion is essentially complete. The Raman spectrum of dinitrogen tetroxide in sulphuric acid[95] clearly shows the nitrosonium ion frequency at 2300 cm^{-1}, and nitrosonium salts are readily prepared from dinitrogen tetroxide.

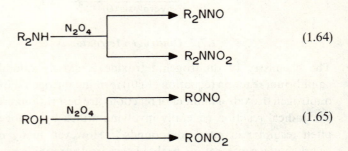

$$\text{(1.61)}$$

Dinitrogen tetroxide has been used to nitrosate a large number of compounds including amines, alcohols and thiols.[96] It appears to be a particularly useful reagent for the deamination of primary amines[97] (equation (1.61)), when fewer by-products are formed than from the more conventional nitrous acid deamination procedure. This reaction is a good synthetic route for the preparation of alkyl and alicyclic nitrates. Further, dinitrogen tetroxide is also an effective nitrosating agent for secondary amides[98] (equation (1.62)), for which other reagents are ineffective. The nitroso–nitrate adduct has also been identified[99] from the reaction of an alkene with dinitrogen tetroxide (equation (1.63)) at low temperature in liquid ethane–propane solvent.

$$RCONHR' + N_2O_4 \xrightarrow{\text{carbon tetrachloride}} RCON(NO)R' \qquad \text{(1.62)}$$

$$CH_2 = C(CH_3)_2 \xrightarrow{N_2O_4} CH_2(NO)C(CH_3)_2ONO_2 \qquad \text{(1.63)}$$

$$R_2NH \xrightarrow{N_2O_4} \begin{cases} R_2NNO \\ R_2NNO_2 \end{cases} \qquad \text{(1.64)}$$

$$ROH \xrightarrow{N_2O_4} \begin{cases} RONO \\ RONO_2 \end{cases} \qquad \text{(1.65)}$$

Nitro compounds are also formed from dinitrogen tetroxide, sometimes alongside the nitroso products. In dichloromethane or acetonitrile solvents secondary amines give both *N*-nitrosamines and *N*-nitroamines,[100,91,26] and alcohols yield alkyl nitrites and alkyl nitrates (see equations (1.64) and (1.65)). The relative yields depend both upon the solvent and the temperature. In general nitrosation is favoured in solvents other than weakly basic solvents such as ethers, and also at higher temperatures.

It has been suggested[91] that nitrosation by dinitrogen tetroxide (other than when there is substantial conversion to NO$^+$NO$_3^-$ in acid solvents) involves another isomeric form, ONONO$_2$, for which there is some evidence from ir spectra.

1.7.3 Nitric oxide NO

There are many literature reports of nitrosations brought about by nitric oxide. However, it is likely that the reactive species is either dinitrogen trioxide or dinitrogen tetroxide formed after aerial oxidation of nitric oxide. When rigorous precautions are taken to exclude all traces of oxygen, reaction of nitric oxide with, for example, amines in acetonitrile solvent is extremely slow, and probably arises from slow diffusion of air into the reaction vessel,[26,91] An exception to this is the reaction of nitric oxide with thiols in basic media discussed in chapter 7.

However, in the presence of metal and other catalysts, nitric oxide can become an effective nitrosating agent. The work of Brackman and Smit[101] showed that the catalysis of the nitrosation of diethylamine by nitric oxide and copper (II) salts was due to the intermediacy of a copper–nitrosyl complex (see scheme (1.66)). A number of metal–nitrosyl complexes show considerable reactivity as nitrosating agents. These are discussed fully in chapter 8.

$$CuCl_2(aq) + NO \rightleftharpoons Cu^{II}NO \text{ complex}$$

$$Cu^{II}NO \text{ complex} + R_2NH \longrightarrow Cu^{I} \text{ complex} + R_2NNO + H^+ \quad (1.66)$$

On the other hand the catalysis by silver ion[102] is thought to derive from oxidation of the amine to a cation radical, which then forms the nitrosamine by reaction with nitric oxide (see (1.67)).

$$\underset{/}{\overset{\backslash}{N}}-H \xrightarrow{Ag^+} \underset{/}{\overset{\backslash}{N}}\overset{+\bullet}{-}H \xrightarrow{NO} \underset{/}{\overset{\backslash}{N}}-NO + H^+ \quad (1.67)$$

Again, when air is rigorously excluded and iodine added, rapid reaction occurs between nitric oxide and secondary amines.[103] This has been interpreted in terms of the intermediate formation of nitrosyl iodide, a well-known nitrosating agent (see equation (1.68)). Since a number of amines appeared to react at about the same rate, the results suggest that the formation of nitrosyl iodide is rate-limiting.

$$NO + \tfrac{1}{2}I_2 \xrightleftharpoons{slow} INO \xrightarrow[\text{rapid}]{R_2NH} R_2NNO \quad (1.68)$$

Nitric oxide can form stable 1:1 complexes with amines,[104] the so-called 'Drago complexes' (equation (1.69)). No detailed structure analysis is available, but it is believed that the anion contains the group:

$$-N\underset{\backslash O^-}{\overset{\diagup N=O}{}}$$

One of the reactions of such a complex is its aerial oxidation. The two main products (equation (1.70)) were identified as *N*-nitrosodiethylamine and diethylammonium nitrite.[105] No mechanistic details have been established.

$$2R_2NH + 2NO = R_2\overset{+}{N}H_2R_2\overset{-}{N}N_2O_2 \qquad (1.69)$$

$$2R_2\overset{+}{N}H_2R_2\overset{-}{N}N_2O_2 + O_2 = 2R_2\overset{+}{N}H_2NO_2^- + 2R_2NNO \qquad (1.70)$$

Since nitric oxide is itself a stable free radical the possibility arises for the formation of nitroso compounds by reaction with other eg carbon radicals. Such a reaction with the triphenylmethyl radical has been long known[106] (equation (1.71)). The product is blue and the reaction reversible, since the reactants can be recovered by evaporation of the solvent (ether). Similarly *C*-nitroso compounds can be generated from nitric oxide if free radicals are generated *in situ* in the reaction medium by the usual procedures of photolysis, pyrolysis or by addition of a stable free radical such as nitrogen dioxide. A good example is the formation of the oxime of cyclohexane (equation (1.72)) via the nitroso compound by a photochemical reaction with chlorine which is believed to generate the cyclohexyl radical.[107] Similarly nitroso alkenes may be formed[108] (equations (1.73) and (1.74)). As expected from radical reactions, some side reactions also occur, so that nitric oxide reactions of this type are not particularly suited for synthesis of nitroso compounds.

$$(C_6H_5)_3C^\cdot + NO \underset{}{\overset{}{\rightleftarrows}} \underset{\text{Blue}}{(C_6H_5)_3CNO} \qquad (1.71)$$

$$(1.72)$$

$$CF_2{=}CFI \xrightarrow{h\nu} CF_2{=}\overset{\cdot}{C}F \xrightarrow{NO} CF_2{=}CF(NO) \qquad (1.73)$$

$$CF_2{=}CHCH_2I \xrightarrow{h\nu} CF_2{=}CH\overset{\cdot}{C}H_2 \xrightarrow{NO} CF_2{=}CHCH_2NO \qquad (1.74)$$

An interesting reaction recently reported[109] involves the catalytic nitrosation of styrene by nitric oxide in dimethylformamide in the presence of a cobalt complex and borohydride (equation (1.75)).

$$PhCH{=}CH_2 + NO \xrightarrow[\text{cobalt complex}]{BH_4^-} \underset{\underset{\;\;\;\;OH}{\overset{||}{N}}}{PhCCH_3} \qquad (1.75)$$

1.8 Miscellaneous reagents

Other well-known nitrosating agents include alkyl nitrites, thionitrites, nitrosamines and metal nitrosyl complexes. These are treated separately in later chapters (chapters 6, 7, 5 and 8 respectively). In addition there have been reports of nitrosations brought about by other reagents; these reactions have, in general, not been much studied mechanistically, but will be included in this chapter.

1.8.1 Nitrite ion

Nitrite ion and nitrous acid (in the absence of added mineral acid) are not normally nitrosating reagents. However, in the presence of some carbonyl compounds, nitrosation of secondary amines can be achieved by nitrite ion in basic and neutral solution. The catalysts include formaldehyde, chloral, benzaldehyde derivatives and pyridoxal. The reaction was discovered by Keefer and Roller,[110] who suggested that the aldehyde and secondary amine formed an iminium ion intermediate, which reacts with nitrite ion forming the dialkylamino nitrite ester which breaks down rapidly to give the dialkyl-nitrosamine (scheme (1.76)). This mechanism has also been suggested to account for nitrosamine formation from secondary amines and solid sodium nitrite in halogenated solvents.[111] The reactions involving form ldehyde and dimethylamine, diethylamine and morpholine have been the subject of a detailed kinetic study.[112] Again reaction is thought to occur by attack of the nitrite ion on the iminium, which itself is formed by dehydration of the protonated form of the carbinolamine, the initial product from the amine and the hydrate of the aldehyde (see scheme (1.77)).

$$R_2NH + R'CHO \xrightleftharpoons[]{OH^-} R_2\overset{+}{N}{=}CHR'$$

$$\downarrow NO_2^-$$

$$(1.76)$$

$$R_2NNO + R'CHO \longleftarrow R_2N{-}CHR'$$

$$\underset{O}{\overset{}{\|}}N{-}O$$

$$R_2NH + H_2C(OH)_2 \rightleftharpoons R_2NCH_2OH + H_2O \Big\}$$

$$R_2NCH_2OH + H^+ \xrightarrow{-H_2O} R_2\overset{+}{N}{=}CH_2 \Big\}$$

$$(1.77)$$

1.8.2 Fremy's salt

Potassium nitrosodisulphonate ($K_2[(SO_3)_2NO]$) was first prepared by

Fremy in 1845.[113] The yellow solid contains the dimeric anion, whilst aqueous solutions, which are violet-blue contain the monomeric anion which is a relatively stable free radical. Fremy's salt has been used as a rather specific (one-electron transfer) oxidising agent in organic chemistry, particularly for the oxidation of substituted phenols and anilines to form quinones.[114] However,[115] it has been shown that secondary and tertiary amines are nitrosated when treated with Fremy's salt in aqueous sodium carbonate solution, or in pyridine solution. In both cases nitrosamine products are formed, the tertiary amine reaction involving N—C bond cleavage and the loss of an alkyl group as an aldehyde. The same reaction occurs using acidified nitrous acid, and involves the iminium cation intermediate (see chapter 4). The same intermediate could be generated here by electron transfer and proton loss (see (1.78)). Reactions are much faster in the presence of added hydroxylamine[116] and high yields of nitrosamines are obtained, which suggests that Fremy's salt oxidises hydroxylamine (added or formed slowly by decomposition of Fremy's salt) to form a species which acts as the final nitrosating agent.

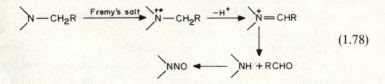

$$(1.78)$$

1.8.3. *Nitro compounds*

There are a number of reports of nitrosations (and nitrations) brought about by aliphatic *C*-nitro compounds. The best-known of these is tetranitromethane $C(NO_2)_4$. *N*, *N*-Dimethylaniline is converted to *N*-methyl-*N*-nitrosoaniline by heating with tetranitromethane in pyridine,[117] (equation (1.79)). Similarly, morpholine is converted to *N*-nitrosomorpholine in good yield by the same reagent in tetrahydrofuran at 70 °C in four hours.[118] Nothing is known about the mechanism, but it is possible that nitrogen dioxide is generated from the nitro compound, forming dinitrogen tetroxide which effects nitrosation.

$$(1.79)$$

Similarly 2-bromo-2-nitropropane 1, 3-diol (an antibacterial and antifungal agent) converts diethanolamine to the nitrosamine[119] (see equation (1.80)). The maximum rate occurs at pH 12 and it has been suggested that nitrite ion is first formed but no details of how nitrosation occurs are known.

$$(CH_2OH)_2C(Br)NO_2 + (HOCH_2CH_2)_2NH \longrightarrow (HOCH_2CH_2)_2NNO$$

$$(1.80)$$

1.8.4 Inorganic nitrates

Again there are a few reports of nitroso compounds being generated from inorganic nitrates. Nitration by nitric acid is also sometimes accompanied by the formation of nitroso compounds. One reagent which has much synthetic potential is 'clayfen', which is a K-10 clay-supported ferric nitrate. This reagent readily converts alcohols stereospecifically into alkyl nitrites at room temperature using a hydrocarbon solvent[120] (equation (1.81)). Similarly thiols are smoothly transformed into thionitrites, though the final isolated products are the disulphides formed by decomposition of the thionitrites and elimination of nitric oxide.[121]

$$ROH \xrightarrow[25\,°C]{clayfen} RONO \qquad (1.81)$$

1.8.5 Sulphur–nitroso compounds

A number of sulphur–nitroso compounds are very reactive nitrosating species. These include nitrososulphinates or sulphonyl nitrites RSO_2NO discussed in chapter 7. The structurally related thionylchloronitrite $SOClONO$ and thionyl dinitrite $SO(ONO)_2$ have recently been prepared from thionyl chloride and silver nitrite in dry tetrahydrofuran.[122] With one equivalent of silver nitrite, thionylchloro nitrite is formed, whereas a second equivalent of silver nitrite yields thionyl dinitrite. Both reagents gave excellent yields of alkyl nitrites from alcohols and oximes from carbonyl compounds.

In section 7.6 the nitrosation of thiosulphate ion is discussed. The product is believed to be[18] the nitrosothiosulphate ion $S_2O_3NO^-$. It has been shown[66] that quite marked catalysis of nitrosation of N-methylaniline occurs by added thiosulphate ion. This suggests strongly that the nitrosothiosulphate ion, although negatively charged, acts as an electrophilic nitrosating agent. A kinetic analysis reveals that it is *ca* 10^4 less reactive than is nitrosyl thiocyanate in the same reaction. The substantial catalysis arises largely because of the very large equilibrium constant for the formation of nitrosothiosulphate ion. Other substrates react similarly.

In the same way there is substantial catalysis by dimethyl sulphide of the nitrosation of a number of compounds.[123] This implies that the nitrososulphonium ion $(CH_3)_2\overset{+}{S}NO$, or something derived from it, has a significant role to play as a nitrosating agent.

Thionitrates $(RSNO_2)$ are also good nitrosating agents in neutral aprotic solvents.[124] Thus arylamines in the presence of copper(II) halides give excellent yields of aryl halides, and secondary amines give nitrosamines. Reaction may occur by rearrangement of the thionitrate to the isomeric nitroso form RS(O)NO. The reactions of thionitrites (RSNO) are discussed in section 7.2.

References

1. K. Jones in *Comprehensive Inorganic Chemistry*, Vol. 2, Eds. J.C. Bailar, H.J. Emeleus, R. Nyholm and A.F. Trotman-Dickenson, Pergamon Press, New York, 1973, p. 368.
2. M. Schumann, *Chem. Ber.*, 1900, **33**, 527.
3. J. Tummavuori & P. Lumme, *Acta Chem. Scand.*, 1968, **22**, 2003; P. Lumme, P. Lahermo & J. Tummavuori, *ibid.*, 1965, **19**, 9; P. Lumme & J. Tummavuori, *ibid.*, 1965, **19**, 617.
4. G. Stedman, *Adv. Inorg. Chem. Radiochem.*, 1979, **22**, 113.
5. G.Y. Markovits, S.E. Schwartz & L. Newman, *Inorg. Chem.*, 1981, **20**, 445.
6. (*a*) K. Singer & P.A. Vamplew, *J. Chem. Soc.*, 1956, 3971; (*b*) N.S. Bayliss & D.W. Watts, *Austral. J. Chem.*, 1956, **9**, 319.
7. W.R. Angus & A.H. Leckie, *Proc. Roy. Soc.*, 1935, **A150**, 615.
8. J.H. Ridd, *Quart. Rev.*, 1961, **15**, 418.
9. L.P. Hammett, *Physical Organic Chemistry*, McGraw-Hill Inc., New York, 1940, p. 294.
10. J. Kenner, *Chem. and Ind.*, 1941, **19**, 443.
11. E.D. Hughes, C.K. Ingold & J.H. Ridd, *J. Chem. Soc.*, 1958, 65, 88.
12. E. Kalatzis & J.H. Ridd, *J. Chem. Soc.* (*B*), 1966, 529.
13. B.C. Challis & A.J. Lawson, *J. Chem Soc., Perkin Trans. 2*, 1973, 918.
14. H. Dahn, L. Loewe & C.A. Bunton, *Helv. Chim. Acta*, 1960, **42**, 320.
15. G. Stedman, *J. Chem. Soc.*, 1959, 2943, 2949.
16. C. Döring & H. Gehlen, *Z. Anorg. Allg. Chem.*, 1961, **312**, 32.
17. J. Casado, A. Castro & M.A. Lopez Quintela, *Monatsh. Chem.*, 1981, **112**, 1221.
18. M.S. Garley & G. Stedman, *J. Inorg. Nucl. Chem.*, 1981, **43**, 2863.
19. C.A. Bunton, D.R. Llewellyn & G. Stedman, *Chem. Soc. Special Publ.*, 1957, **10**, 113; *J. Chem. Soc.*, 1959, 568; C.A. Bunton & G. Stedman, *J. Chem. Soc.*, 1959, 3466.
20. J. Casado, A. Castro, J.R. Leis, M.A. Lopez Quintela & M. Mosquera, *Monatsh. Chem.*, 1983, **114**, 639.
21. J.R. Park & D.L.H. Williams, *J. Chem. Soc., Perkin Trans. 2*, 1972, 2158.
22. L. Greci & J.H. Ridd, *J. Chem. Soc., Perkin Trans. 2*, 1979, 312.

23. A.M.M. Doherty, M.S. Garley, K.R. Howes & G. Stedman, *J. Chem. Soc., Perkin Trans. 2*, 1986, 143.
24. B.C. Challis & A.R. Butler in *The Chemistry of the Amino Group*, Ed. S. Patai, Wiley-Interscience, London 1968, Ch. 6, p. 305.
25. J.H. Ridd, *Adv. Phys. Org. Chem.*, 1978, **16**, 1.
26. B.C. Challis & S.A. Kyrtopoulos, *J. Chem. Soc., Perkin Trans. 1*, 1979, 299.
27. E.D. Hughes, C.K. Ingold & J.H. Ridd, *J. Chem. Soc.*, 1958, 77.
28. E.D. Hughes, C.K. Ingold & J.H. Ridd, *J. Chem Soc.*, 1958, 88.
29. 'Stability Constants', *Chem. Soc., Special Publ.*, 1964, **17**, 164.
30. B.C. Challis & J.H. Ridd, *Proc. Chem. Soc.*, 1960, 245.
31. D.J. Benton & P. Moore, *J. Chem. Soc. (A)*, 1970, 3179.
32. J.O. Edwards, *Science*, 1951, **113**, 392.
33. N.C. Marziano, J.H. Rees & J.H. Ridd, *J. Chem. Soc., Perkin Trans. 2*, 1974, 600.
34. L.R. Dix & D.L.H. Williams, *J. Chem. Res. (S)*, 1982, 190.
35. F. Seel & R. Winkler, *Z. Phys. Chem.*, 1960, **25**, 217; H. Schmid & P. Krenmayr, *Monatsh. Chem.*, 1967, **98**, 417.
36. (*a*) M.T. Nguyen & A.F. Hegarty, *J. Chem. Soc., Perkin Trans. 2*, 1984, 2037;
    (*b*) K.A. Jørgensen & S.O. Lawesson, *ibid.*, 1985, 231.
37. J. Fitzpatrick, T.A. Meyer, M.E. O'Neill & D.L.H. Williams, *J. Chem. Soc., Perkin Trans. 2*, 1984, 927.
38. S.S. Mirvish, *Toxic. Appl. Pharmacol.*, 1975, **31**, 325.
39. J.R. Perrott, G. Stedman & N. Usyal, *J. Chem. Soc., Dalton Trans.*, 1976, 2058.
40. S.E. Aldred, D.L.H. Williams & M. Garley, *J. Chem. Soc., Perkin Trans. 2*, 1982, 777.
41. J.C.M. Li & D.M. Ritter, *J. Am. Chem Soc.*, 1953, **75**, 3024.
42. H. Schmid, *Chem.-Ztg.*, 1962, **86**, 811.
43. P. Collings, K. Al-Mallah & G. Stedman, *J. Chem. Soc., Perkin Trans. 2*, 1975, 1734.
44. L.R. Dix & D.L.H. Williams, *J. Chem. Soc., Perkin Trans. 2*, 1984, 109.
45. E.D. Hughes & J.H. Ridd, *J. Chem Soc.*, 1958, 82.
46. G. Stedman, *J. Chem. Soc.*, 1960, 1702.
47. T. Bryant & D.L.H. Williams, *J. Chem. Soc., Perkin Trans. 2*, 1985, 1083.
48. L.J. Beckman, W.A. Fessler & M.A. Kise, *Chem. Rev.*, 1951, **48**, 319.
49. B.C. Challis & D.E.G. Shuker, *J. Chem. Soc., Perkin Trans 2*, 1979, 1020.
50. Reference 1 pp. 303, 311.
51. M.P. Doyle, R.J. Bosch & P.G. Seites, *J. Org. Chem.*, 1978, **43**, 4120.
52. R. Weiss & K.G. Wagner, *Chem. Ber.*, 1984, **117**, 1973.
53. J.H. Boyer in *The Chemistry of the Nitro and Nitroso Groups* Ed. H. Feuer, Wiley-Interscience, New York 1969, p. 220.
54. H. Schmid & E. Hallaba, *Monatsh. Chem.*, 1956, **87**, 560; H. Schmid and M.G. Fouad, *ibid.*, 1957, **88**, 631.
55. A. Woppmann & H. Sofer, *Monatsh. Chem.*, 1972, **103**, 163.
56. H. Schmid, *Z. Electrochem.*, 1937, **43**, 626; H. Schmid & G. Muhr, *Chem. Ber.*, 1937, **70**, 421.

57. P.A. Morris & D.L.H. Williams to be published.
58. D.L.H. Williams, *Chem. Soc. Rev.*, 1985, **14**, 171.
59. J. Casado, J.R. Gallastegui, M. Losada, L.C. Paz & J.V. Tato, *Acta Cient. Comp.*, 1982, **19**, 209.
60. H. Schmid, *Monatsh. Chem.*, 1954, **85**, 424.
61. M.R. Crampton, J.T. Thompson & D.L.H. Williams, *J. Chem. Soc., Perkin Trans. 2*, 1979, 18.
62. H. Schmid & C.H. Essler, *Monatsh. Chem.*, 1957, **88**, 1110; H. Schmid & M.G. Fouad, *ibid.*, p. 631.
63. T.D.B. Morgan, G. Stedman & M.N. Hughes, *J. Chem. Soc. (B)* 1968. 344.
64. B.C. Challis & R.J. Higgins, *J. Chem. Soc., Perkin Trans. 2*, 1975, 1498.
65. T.Y. Fan & S.R. Tannenbaum, *J. Agr. Food Chem.*, 1973, **21**, 237.
66. T. Bryant, D.L.H. Williams, M.H.H. Ali & G. Stedman, *J. Chem. Soc., Perkin Trans. 2*, 1986, 193.
67. S.E. Aldred & D.L.H. Williams, *J. Chem. Soc., Perkin Trans. 2*, 1981, 1021.
68. K.A. Jørgensen & S.O. Lawesson, *J. Am. Chem. Soc.*, 1984, **106**, 4687.
69. C.C. Addison & J. Lewis, *Quart. Rev.*, 1955, **9**, 115.
70. G. Stedman & P.A.E. Whincup, *J. Chem. Soc.*, 1963, 5796.
71. D.L.H. Williams, *Adv. Phys. Org. Chem.*, 1983, **19**, 381.
72. E. Boyland & S.A. Walker, *Nature (London)* 1974, **248**, 601.
73. J. Casado, A. Castro, M.A. Lopez Quintela & M.F.R. Prieto, *Monatsh. Chem.*, 1983, **114**, 647; J. Casado, A. Castro, M. Mosquera, M.F.R. Prieto & J.V. Tato, *ibid.*, 1984, **115**, 669.
74. J. Hohle & F.C. Mijlhoff, *Recl. Trav. Chim. Pays-Bas*, 1967, **86**, 1153.
75. D.W.A. Sharp & J. Thorely, *J. Chem. Soc.*, 1963, 3557.
76. F. Seel & T. Gössl, *Z. Anorg. Allg. Chem.*, 1950, **263**, 253.
77. U. Wannagat & G. Hohlstein, *Chem. Ber.*, 1955, **88**, 1839; G.A. Olah & W.S. Tolgyesi, *J. Org. Chem.*, 1961, **26**, 2319.
78. G.A. Olah & J.A. Olah in *Friedel Crafts and Related Reactions*, Vol. 3, Ed. G.A. Olah Interscience, New York, 1964, 1267.
79. J.M. Simpson, D.C. Kapp & T.M. Chapman, *Synthesis*, 1979, 100.
80. G.A. Olah, N.A. Overchuk & J.C. Lapièrre, *J. Am. Chem. Soc.*, 1965, **87**, 5785.
81. F. Seel & H. Sauer, *Z. Anorg. Allg. Chem.*, 1957, **292**, 1 and earlier papers.
82. G.A. Olah, *Acc. Chem. Res.*, 1980, **13**, 330.
83. (a). B.K. Bandlish & H.J. Shine, *J. Org. Chem.*, 1977, **42**, 561;
    (b). W.K. Musker, T.L. Wolford & P.B. Roush, *J. Am. Chem. Soc.*, 1978, **100**, 6416; W.K. Musker, A.S. Hirschon & J.T. Doi, *ibid.*, p. 7754.
84. A.N. Koshechko, A.N. Inozemtsev & V.D. Pokhodenko, *Zh. Org. Khim.*, 1983, **19**, 751.
85. K. Al-Mallah, P. Collings & G. Stedman, *J. Chem. Soc., Dalton Trans.*, 1974, 2469.
86. T.A. Meyer & D.L.H. Williams, *J. Chem. Soc., Perkin Trans. 2*, 1981, 361.
87. M. Masui, C. Ueda, T. Yasuoka & H. Ohmori, *Chem. Pharm. Bull.*, 1979, **27**, 1274.
88. Reference 1, pp. 335–40.

89. D.J. Lovejoy & A.J. Vosper, *J. Chem. Soc. A*, 1968, 2325.
90. O. Touster, *Org. Reactions*, 1953, **7**, 327.
91. B.C. Challis & S.A. Kyrtopoulos, *J. Chem. Soc., Perkin Trans. 2*, 1978. 1296.
92. H. Shechter, *Rec. Chem. Progr.*, 1964, **25**, 55.
93. Reference 1, pp. 340–56.
94. C.C. Addison, *Angew. Chem.*, 1960, **72**, 193; *Chem. Rev.*, 1980, **80**, 21.
95. D.J. Millen, *J. Chem. Soc.*, 1950, 2589.
96. P. Gray & A.D. Yoffe, *Chem. Rev.*, 1955, **55**, 1069.
97. D.H.R. Barton & S.C. Narang, *J. Chem. Soc., Perkin Trans. 1*, 1977, 1114.
98. E.H. White, *J. Am. Chem. Soc.*, 1955, **77**, 6008.
99. L. Parts & J.T. Miller, *J. Phys. Chem.*, 1969, **73**, 3088.
100. E.H. White & W.R. Feldman, *J. Am. Chem. Soc.*, 1957, **79**, 5832.
101. W. Brackman & P.J. Smit, *Recl. Trav. Chim. Pays-Bas*, 1965, **84**, 357, 372.
102. B.C. Challis & J.R. Outram, *J. Chem. Soc., Chem. Commun.*, 1978, 707.
103. B.C. Challis & J.R. Outram, *J. Chem. Soc., Perkin Trans. 1*, 1979, 2768.
104. R.S. Drago & F.E. Paulik, *J. Am. Chem. Soc.*, 1960, **82**, 96.
105. R.O. Ragsdale, B.R. Karstetter & R.S. Drago, *Inorg. Chem.*, 1965, **4**, 420.
106. W. Schlenk, L. Mair & C. Bornhardt, *Chem. Ber.*, 1911, **44**, 1169.
107. H. Metzger & E. Müller, *Chem. Ber.*, 1957, **90**, 1179.
108. C.E. Griffen & R.N. Haszeldine, *Proc. Chem. Soc.*, 1959, 369; P. Tarrant & J. Savory, *J. Org. Chem.*, 1963, **28**, 1728.
109. T. Okamoto & S. Oka, *J. Chem. Soc., Chem. Commun.*, 1984, 289.
110. L.K. Keefer & P.P. Roller, *Science*, 1973, **181**, 1245.
111. P.P. Roller, L.K. Keefer & B.W. Slavin in *N-Nitroso Compounds: Analysis, Formation and Occurrence*, Eds. E.A. Walker, L. Gricuite, M. Castegnaro and M. Brozonyi, IARC Scientific Publication 31, Lyon, 1980, 119.
112. J. Casado, A. Castro, M.A. Lopez Quintela & J.V. Tato, *Z. Phys. Chem., Neue Folge*, 1981, **127**, 179; J. Casado, M. Mosquera, L.C. Paz, M.F.R. Prieto & J.V. Tato, *J. Chem. Soc., Perkin Trans. 2*, 1984, 1963.
113. H. Fremy, *Ann. Chim. Phys.*, 1845, **15**, 408.
114. H. Zimmer, D.C. Lankin & S.W. Horgan, *Chem. Rev.*, 1971, **71**, 229.
115. L. Castedo, R. Riguera & M.P. Vazquez, *J. Chem. Soc., Chem. Commun.*, 1983, 301.
116. M.P.V. Tato, L. Castedo & R. Riguera, *Chem. Lett.*, 1985, 623.
117. E. Schmidt & R. Schumacker, *Chem. Ber.*, 1921, **54**, 1414.
118. T.Y. Fan, R. Vita & D.H. Fine, *Toxicol. Lett.*, 1978, **2**, 5.
119. I. Schmeltz & A. Wenger, *Food Cosmet. Toxicol.*, 1979, **17**, 105.
120. A. Cornelius, P.Y. Herze & P. Laszlo, *Tetrahedron Lett.*, 1982, **23**, 5035.
121. A. Cornelius, N. Depaye, A. Gerstmans & P. Laszlo, *Tetrahedron Lett.*, 1983, **24**, 3103.
122. G.H. Hakimelahi, H. Sharghi, H. Zarrinmayeh & A. Khalafi-Nezhad, *Helv. Chim. Acta*, 1984, **67**, 906.
123. T. Bryant & D.L.H. Williams, *J. Chem. Res(S)*, 1987, 174.
124. S. Oae & K. Shinhama, *Org. Prep. Proced. Int.*, 1983, **15**, 165.

2

Aliphatic and alicyclic C-nitrosation

Aliphatic and alicyclic *C*-nitroso compounds can exist as monomers, dimers or as the tautomeric oximes. In general the dimers and the oximes are the more stable forms, but in some cases structural features stabilise the monomer. When the monomer is stable, it is easily recognised by its blue or green colour due to a n → π* transition which occurs in the range 630 – 790 nm, with low extinction coefficients ($\varepsilon = 1$–60 mol^{-1} dm^3 cm^{-1}). The ir N—O bond stretching frequency is also reasonably diagnostic, occurring in the range 1540–1620 cm^{-1} in the monomer.

The diazene 1, 2-dioxide structure (see equation (2.1)) of the dimer is now well established, principally by ir spectroscopy and by X-ray crystal structure analyses. Bonding undoubtedly occurs between the nitrogen atoms but resonance forms other than those shown in equation (2.1) are involved, since the N—N bond order is about 1.5. Both the Z and E forms are known, with the latter being the more stable, although generally the Z isomer is the first formed. In aliphatic E dimers the N—O bond stretching frequency occurs at 1176–1290 cm^{-1}. The dimers are colourless as solids and in solution. They also melt to give initially colourless liquids, but on further heating dissociate to the coloured monomers. Dissociation can also occur in solution. The activation barrier for dissociation is quite large at 80–120 kJ mol^{-1}, and so is not necessarily a rapid process; dimerisation has a lower energy barrier at *ca* 25–40 kJ mol^{-1}. Tertiary nitroso compounds often exist as monomers, because of the steric strain of dimerisation. The presence of electron-withdrawing groups also stabilises the monomer.

$$2RNO \; \rightleftharpoons \quad \underset{^-O}{\overset{R}{}}\!\! \overset{+}{N}\!\!=\!\!\overset{+}{N}\!\! \underset{R}{\overset{O^-}{}} \quad + \quad \underset{^-O}{\overset{R}{}}\!\! \overset{+}{N}\!\!=\!\!\overset{+}{N}\!\! \underset{O^-}{\overset{R}{}} \tag{2.1}$$

The formation of oximes from both primary and secondary nitroso compounds (equations (2.2)) is one of the principal synthetic methods used to generate C═N structures.[1] This isomerisation has a significant energy barrier (*ca* 150 kJ mol^{-1} for gas phase reactions), and is also both acid- and base-catalysed in solution. Further the reaction is also solvent and temperature dependent, so the isolated products of *C*-nitrosation of aliphatic and alicyclic compounds may be nitroso monomers, nitroso dimers, or oximes, depending on the experimental conditions. Often the product remains in the nitroso monomer form if the nitrosation is carried out in the absence of acids or bases, ie using nitrogen oxides such as dinitrogen trioxide.

$$\left.\begin{array}{c} RCH_2NO \rightleftharpoons RCH{=}NOH \\ RCH(X)NO \rightleftharpoons RCX{=}NOH \end{array}\right\} \qquad (2.2)$$

The physical properties and chemical reactions of *C*-nitroso compounds are well documented in three review articles.[2,3,4] The remainder of this chapter will be concerned only with the formation of *C*-nitroso compounds by nitrosation reactions.

2.1 Reactions of ketones

It has long been known that compounds containing methyl or methylene groups adjacent to a carbonyl group, are readily nitrosated to give usually the corresponding oximes. This applies also to a range of other electron-withdrawing groups, and the reactants are sometimes said to possess an 'active' methyl or methylene group. In the early literature the products were often referred to as 'isonitroso' compounds, but nowadays they are recognised as oximes. The process is typified by the reactions of acetone and acetylacetone shown in equations (2.3) and (2.4). Reaction is quite general for both aliphatic and aromatic monoketones as well as for β-diketones, β-ketoacids, β-ketoesters, haloketones, and other related structures, although in some of these reactions the initial product undergoes a further C—C bond fission, which is discussed more fully later. The reactions of compounds containing other electron-withdrawing groups are discussed in sections 2.2 and 2.3.

$$CH_3COCH_3 \xrightarrow{\text{XNO}} CH_3COCH{=}NOH \qquad (2.3)$$

$$CH_3COCH_2COCH_3 \xrightarrow{\text{XNO}} CH_3COCCOCH_3 \qquad (2.4)$$
$$\underset{\displaystyle NOH}{\overset{\displaystyle \|}{}}$$

The most commonly used nitrosating agents have been aqueous acidic nitrous acid, and alkyl nitrites in ether or ethanol containing hydrogen chloride. Other systems which have been used synthetically include alkyl nitrites and alkoxide ion, nitrosyl sulphuric acid, nitrosyl chloride and nitrogen oxides. Aromatic nitrosamines have also been used to form oximes from 'activated' methylene groups,[5] though it is not known whether this is a direct reaction or whether N—N bond fission first occurs to give some nitrosating species. No doubt any of the reagents described in chapter 1 could be effective. A comprehensive survey has been prepared[6] of nitrosations at aliphatic carbon (up to 1953), including the practical details of some of them.

With unsymmetrical ketones such as methyl alkyl ketones (equation (2.5)) and chloroacetone (equation (2.6)) the normal product is the oxime arising from nitrosation in the methylene rather than the methyl group.

$$RCH_2COCH_3 \xrightarrow{XNO} \underset{\substack{\| \\ NOH}}{RCCOCH_3} \tag{2.5}$$

$$ClCH_2COCH_3 \xrightarrow{XNO} \underset{\substack{\| \\ NOH}}{ClCCOCH_3} \tag{2.6}$$

Cyclic ketones form the α, α'-disubstituted oximes[6] (see equation (2.7)). It does not seem possible to stop the reaction at the monooxime stage. This is a major difference between cyclic ketones (at least for C(5)–C(8)) and open-chain ketones where mononitrosation is the rule. An explanation has been advanced,[7] based on the relative ease with which the proposed intermediate nitroso carbocations isomerise to the oxime, in the open-chain and cyclic structures.

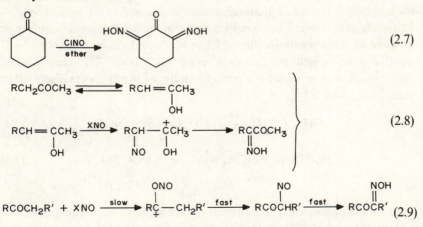

Very little is known about the detailed mechanism of aliphatic *C*-nitrosation. This is rather surprising, given the synthetic importance of the reactions, and contrasts markedly with *N*-nitrosation (chapter 4), where considerable effort has been directed successfully at establishing detailed reaction pathways. There is a fairly general belief, which until very recently has not been backed up by experiment, that reaction involves enolisation of the ketone (or more generally tautomerisation), which is followed by electrophilic nitrosation of the enol (or related structure), as outlined in scheme (2.8). This is analogous to the familiar halogenations, deuterations and racemisations of ketones. In these latter examples enolisation is usually rate-limiting. One kinetic study[8(a)] however, has claimed that the nitrosation of acetone was significantly faster than the enolisation, under the same conditions, which appears to rule out the mechanism in scheme (2.8). These authors[8(a)] postulated an alternative scheme (scheme (2.9)) in which the electrophile attacks the keto form at the carbonyl oxygen atom in the rate-limiting step, which is followed by a rapid internal NO group rearrangement. Some support for the existence of a nitroso carbocation intermediate comes from a detailed product.analysis of the nitrosation of cyclohexanone.[7] When nitrosyl chloride is used, in a variety of organic solvents, only the dioxime is formed. However, in liquid sulphur dioxide containing one equivalent of methanol and of dry hydrogen chloride at $-70\,°C$, 2-methoxy-3-oximinocyclohexene is the isolated product. This is rationalised by scheme (2.10), where the initially formed nitroso carbocation is captured by the nucleophilic methanol. The 1-hydroxy-1-methoxy-2-oximinocyclohexane hydrochloride was detected by nmr; treatment with base and warming then gave the hexene product. Interestingly the role of hydrogen chloride is crucial, since in its absence, C—C bond fission occurs resulting in the formation of methyl ω-oximinocaproate. This is explained in terms of the reaction of the nitroso hemiacetal given in equation (2.11).

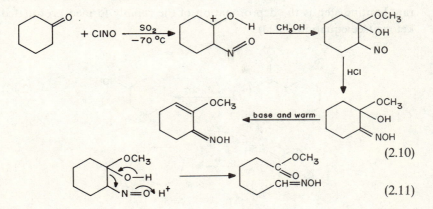

(2.10)

(2.11)

In the absence of efficient nucleophiles, cyclic products have been isolated, which are believed to be formed from the intermediate nitroso carbocation by 1, 3-dipolar addition with the carbonyl group of the reactant. From cyclohexanone itself, with nitrosyl chloride in ether saturated with hydrogen chloride at $-55\,°C$, the isolated product is the oxazoline *N*-oxide shown in equation (2.12).

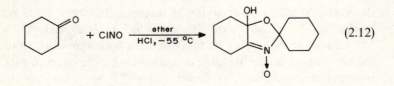

$$\text{(2.12)}$$

C—C bond fissions (sometimes called nitrolyses) are quite well known in the nitrosation of some ketones,[6,9] and of other substrates containing electron-withdrawing groups. This is particularly true if the nitroso group is on a tertiary carbon atom (and hence cannot give the stable oxime), and also when there are two electron-withdrawing groups present. An industrially important example of this, is the formation of the oxime of cyclohexanone from a range of substituted cyclohexane derivatives including aryl cyclohexyl ketones[10] (equation (2.13)) and cycloalkyl carboxylic acids.[11] The oxime then undergoes a Beckmann rearrangement in acid solution to form the seven-membered cyclic amide, ε-caprolactam (see equation (2.14)), which is the monomer used in the production of nylon 6. The nitrosation can be brought about with either nitrosyl chloride or nitrosyl sulphuric acid. Again, not much is known about the reaction mechanism, except that in 85–95% sulphuric acid the rate of reaction decreases with increasing acidity. The kinetic data correlate with the decreasing water activity, and a mechanism is assumed, based on the enolisation of the ketone, where the rate-limiting step is the deprotonation of the rapidly formed protonated ketone (see equation (2.15)).

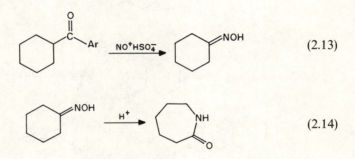

$$\text{(2.13)}$$

$$\text{(2.14)}$$

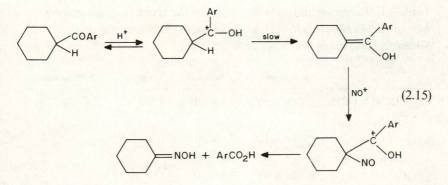

$$(2.15)$$

A later mechanistic study[8(b)], however, has shown clearly that the nitrosation of ketones does indeed involve reaction of the enol form as is outlined in scheme (2.8). Particularly in the presence of nucleophilic species X^-(Cl^-, Br^-, SCN^- and $SC(NH_2)_2$) the reaction is first-order in both [ketone] and [H^+] and zero-order in both [HNO_2] and [X^-]. These results were obtained for the reactions of both acetone and for methyl ethyl ketone in aqueous solution. The situation then very strongly resembles that of the halogenation of ketones and involves under these conditions rate-limiting enolisation. The rate constants obtained for enolisation from the nitrosation experiments agree well with those obtained in the halogenation reactions. By adjusting both [X^-] and [HNO_2] it is possible to change the rate-limiting step from that of enolisation to attack of the enol by XNO. Now the order with respect to both [X^-] and [HNO_2] is one, and analysis of the data enables the second-order rate constants for attack by XNO to be obtained. Under some conditions the corresponding reaction orders are intermediate between one and two, and the results are readily rationalised in terms of scheme (2.8). Some results for the reactions of acetone, methyl ethyl ketone and acetylacetone are shown in table 2.1. In the case of acetylacetone it is not possible to achieve rate-limiting enolisation, no doubt in part due to the much reduced rate constant for attack by XNO occasioned by the electron-withdrawing —$COCH_3$ group. The values in table 2.1 are based on a literature value for the keto–enol equilibrium constant, which for acetone and methyl ethyl ketone is very small and subject to some uncertainty. Nevertheless the results show the expected trend of reactivity nitrosyl chloride > nitrosyl bromide > nitrosyl thiocyanate > *S*-nitrosothiouronium ion which is now well established particularly in *N*-nitrosation. The figures also show that as expected acetone and methyl ethyl ketone are significantly more reactive than is acetylacetone.

Table 2.1. *Rate constants* $(l\,mol^{-1}\,s^{-1})$ *for the reaction of some enols with nitrosyl chloride, nitrosyl bromide, nitrosyl thiocyanate and S-nitrosothiouronium ion in water at 25 °C*

	k_{ClNO}	k_{BrNO}	k_{ONSCN}	$k_{(NH_2)_2C\overset{+}{S}NO}$

CH₃COCH=CCH₃
          |
          OH
	1.0×10^5	1.4×10^4	500	38

CH₂=CCH₃
    |
    OH
	1.4×10^8	7.0×10^7	–	–

CH₃CH=CCH₃
        |
        OH
	4.6×10^9	3.8×10^9	$\sim 3 \times 10^8$	–

In the absence of added nucleophiles the nitrosation reaction of ketones is much slower and uncertainties arise from the decomposition of nitrous acid and now also from the hydrolysis of the oxime product and subsequent nitrosation of the hydroxylamine formed. However, it has been shown[8(b)] that dinitrogen trioxide is the principal reagent in the nitrosation of the enols of acetone and methyl ethyl ketone (and reaction is close to the diffusion limit) whereas for the less reactive acetylacetone derivative reaction occurs via the nitrous acidium ion or the nitrosonium ion.

2.2 Reactions of other carbonyl compounds

A range of compounds other than simple ketones containing the carbonyl group adjacent to a methyl or methylene group, also react readily with a variety of nitrosating agents, generally to give the corresponding oximes. Some examples are given in equations (2.16) and (2.17). Again the C—C bond cleavage reaction occurs with suitable substituents, as for the ketone reactions. Examples are shown in equations (2.18) and (2.19) for the reactions of α-substituted β-keto esters and alkyl malonic acids respectively. Cyclic compounds react similarly. Some of the reactions have been developed as general synthetic routes. The most notable example is the nitrosation of cyclohexanecarboxylic acid, which has been established industrially in the Snia–Viscosa process for the manufacture of ε-caprolactam[12] (equation (2.20)). Again as for the ketone reactions, the usual reagents are nitrosyl sulphuric acid or nitrosyl chloride. The reaction is quite general for other ring sizes, but little is known of the detailed reaction mechanism. It has been suggested[13] that decarboxylation occurs by way of a cyclic mechanism (see equation (2.21)), which is analogous to that belie-ved to occur in the decarboxylation of β,γ-unsaturated carboxylic acids. Another view is that a ketene intermediate is first formed, and that this is the species which is nitrosated (see equation (2.22)). Kinetic studies have been reported,[14] including the catalytic effect of added sulphur trioxide, but they are not diagnostic of a reaction mechanism. There are many reports in the Japanese patent literature which refer to the formation of caprolactam from cyclopentamethylene ketene and nitrosyl sulphuric acid. Again the reaction is quite general for other cycloalkylene ketenes.

$$CH_3COCH_2CO_2C_2H_5 \xrightarrow{\text{XNO}} CH_3COCCO_2C_2H_5 \qquad (2.16)$$
$$\overset{\|}{\underset{\text{NOH}}{}}$$

$$C_6H_5CH_2CO_2C_2H_5 \xrightarrow{\text{XNO}} C_6H_5CCO_2C_2H_5 \qquad (2.17)$$
$$\overset{\|}{\underset{\text{NOH}}{}}$$

$$RCOCH(R')CO_2R'' \xrightarrow{\text{XNO}} R'CCO_2R'' \qquad (2.18)$$
$$\overset{\|}{\underset{\text{NOH}}{}}$$

$$RCH(CO_2H)_2 \xrightarrow{\text{XNO}} RCCO_2H \qquad (2.19)$$
$$\overset{\|}{\underset{\text{NOH}}{}}$$

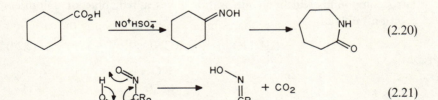

$$(2.20)$$

$$(2.21)$$

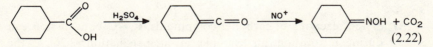

$$(2.22)$$

The industrial process for caprolactam production has used nitrosyl chloride as the reagent irradiated with uv light, when clearly a radical mechanism is involved. These reactions are also carried out in acid solution, which assists the tautomerisation of the nitroso compound to the oxime, and also effects the rearrangement to the cyclic amide.

Malonic esters also react to form the corresponding oximes (equation (2.23)), a reaction which has been used as the first stage of amino acid synthesis. The oxime is reduced and the resulting amine is acetylated. Subsequent alkylation, hydrolysis and decarboxylation gives the amino acid.

$$CH_2(CO_2R)_2 \xrightarrow{XNO} HON=C(CO_2R)_2 \qquad (2.23)$$

2.3 Reactions of compounds containing other electron-withdrawing groups

The reactions outlined in sections 2.1 and 2.2 occur quite generally when any electron-withdrawing group is adjacent to a methyl or methylene group, although most of the literature references are to carbonyl compounds. The first example of such a reaction, however, involved a nitro compound and was reported by Victor Meyer in 1873.[15] Primary nitro compounds form α-nitro oximes (nitrolic acids) as in equation (2.24). The anions of these substances are characteristically red. Secondary nitro compounds (equation (2.25)) form α-nitro nitroso products (pseudonitroles), which are blue (as the monomers) in solution. Tertiary nitro compounds do not react. These reactions are the basis of a colour test devised by Victor Meyer to distinguish generally amongst primary, secondary and tertiary structures. The unknown (alcohol, alkyl halide or amine) is readily converted by test-tube reactions into the corresponding nitro compound, which on nitrosation gives a red, blue or colourless solution in alkali.

$$RCH_2NO_2 \xrightarrow{XNO} \underset{\underset{NOH}{\|}}{RCNO_2} \qquad (2.24)$$

$$RCH(R')NO_2 \xrightarrow{XNO} \underset{\underset{NO}{|}}{RC(R')NO_2} \qquad (2.25)$$

As for the reactions of carbonyl compounds, C—C bond fission can occur during nitrosation; an example, given in equation (2.26), is the reaction of 1,3-dihydroxy-2-nitropropane, where the products are the hydroxy α-nitro oxime and formaldehyde.

$$CH(NO_2)(CH_2OH)_2 \xrightarrow{HNO_2} \underset{\underset{NOH}{\|}}{HOCH_2CNO_2} + HCHO \qquad (2.26)$$

It is reasonable to assume, although there are no mechanistic studies reported, that reaction occurs via the nitronic acid (or the nitronate anion $RCH=NO_2^-$),[16] the tautomeric form of the nitro compound (see scheme (2.27)). This pathway is analogous to that now firmly

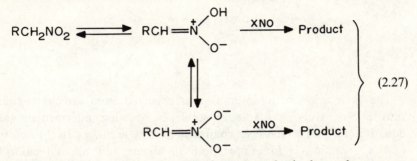

$$(2.27)$$

established for the reactions of carbonyl compounds via the enol tautomer.

Other electron-withdrawing substituents which are known to promote C-nitrosation in this way, include the cyano, imino, halo and aryl groups.

Two less familiar examples are shown in equations (2.28) and (2.29) for the reactions of the pyridiomethyl cobalt (III) ion,[17] and of phosphono (and phosphino) acetic acid esters.[18] Not unexpectedly cyclopentadiene yields the oxime on nitrosation (equation (2.30))[19] reflecting the well-known reactivity of the hydrogen atom of the methylene group in cyclopentadiene, in the absence of a formal electron-withdrawing group.

$$[H\overset{+}{N}C_5H_4CH_2Co(CN)_5]^{3-} \xrightarrow[H^+]{HNO_2} H\overset{+}{N}C_5H_4CH=NOH \quad (2.28)$$

$$(RO)R'P(O)CH_2CO_2R'' \xrightarrow{ClNO} (RO)R'P(O)CCO_2R'' \quad (2.29)$$
$$\underset{NOH}{\|}$$

$$\xrightarrow[RO^-]{RONO} \quad =NOH \quad (2.30)$$

2.4 Additions to alkenes

2.4.1 Nitrosyl halides

The early work of Tilden & Stenstone[20] first demonstrated that nitrosyl chloride reacts with alkenes to give nitroso chloro adducts (equation (2.31)).The products are often (but not always) isolated as the nitroso dimers or as the oximes. In many cases the nitroso chloro compounds are unstable so reaction is often carried out at low temperature (*ca* − 50 °C). In other cases, the products are stable crystalline materials, which have been much used in the past to characterise alkenes; this is particularly true in the chemistry of the terpenes.[21]

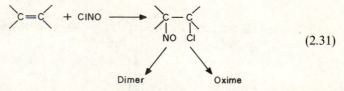

$$(2.31)$$

The orientation of addition is that expected from an electrophilic nitrosation, involving the more stable carbocation intermediate (see equation (2.32)). This has been confirmed by kinetic studies. In chloroform at 0 °C, reaction is first-order both in alkene and nitrosyl chloride concentrations, and the substituent effects (in thirty alkenes studied) support a rate-limiting electrophilic nitrosation.[22] Steric effects have also been shown to be important, and the rate constants increase with the polarity of the solvent.

$$(CH_3)_2C\!=\!CH_2 \xrightarrow{\text{ClNO}} (CH_3)_2\overset{+}{C}\!-\!CH_2NO \longrightarrow \text{Product} \quad (2.32)$$

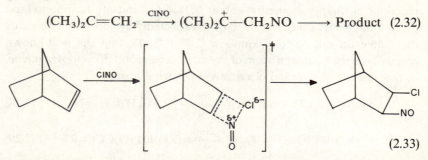

$$(2.33)$$

The stereochemistry of the addition has been established in some cases, although the early literature is rather confused on this topic. Both *syn-* and *anti*-addition have been reported. The reaction mechanism is dependent upon both the alkene structure and also the solvent. For bicyclic systems such as norbornene and norbornadiene[23] *syn*-addition is preferred. This, together with the absence of any skeletal rearrangement during nitrosation and also the absence of any products derived by diverting any intermediate carbocation with an added nucleophile, suggests that very little positive charge is developed on carbon. A mechanism was proposed involving a four-centre transition state as outlined in scheme (2.33), which accounts for the experimental facts. *Syn*-addition also occurs in the reaction of 1, 5-cyclooctadiene with nitrosyl chloride in methylene chloride.[24] Generally, it appears that *syn*-addition predominates with bicyclic systems and also in solvents of low polarity such as methylene chloride and trichloroethane. However, in more polar solvents such as formic acid and liquid sulphur dioxide, products of *anti*-addition of nitrosyl chloride are formed.[25] These results are rationalised in the same way as are other electrophilic additions to alkenes which show *anti*-addition, eg halogenation generally.[26] It seems that a two-stage process involving a nitroso carbocation can take place in a solvent which is sufficiently polar to stabilise such intermediates and the transition states leading to them. The stereochemistry is accounted for by a cyclic onium structure involving either a fully-bonded three-membered ring, or an electrostatic interaction between the nitrogen atom and the developing positive charge on carbon (see equation (2.34)).

$$ (2.34) $$

A more detailed study of the stereochemical course of the reaction of nitrosyl chloride with cyclohexene has been carried out.[27] A pair of enantiomeric *trans*(1S, 2S)- and *trans*(1R, 2R)-2-chloronitrosocyclohexanes are formed. Dimerisation can then occur with two like- or unlike-enantiomers. This is called homospecific and heterospecific dimerisation, leading to two different isolable dimeric chloro nitroso products.

As expected enol ethers are very reactive towards nitrosyl chloride. The isolated product from a cyclohexanone enol is in fact a cyclohexanone oxime (see equation (2.35)) which is probably formed by elimination of hydrogen chloride from the nitroso chloro adduct. If the nitroso group is tertiary, then the initial product is also unstable to spontaneous C—C bond cleavage,[28] as outlined in equation (2.36). An interesting synthetic application (which is of pharmaceutical interest) involves the C—C bond cleavage reaction in cyclic enol ethers,[29] to give initially oximino macrolides, which are readily hydrolysed to give the keto macrolides (equation (2.37)).

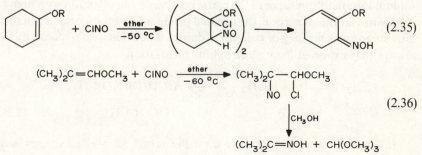

$$ (2.35) $$

$$ (2.36) $$

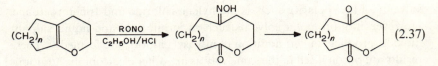

Good yields of nitroso chloro products have been obtained[30] using nitrosyl chloride generated *in situ* from hydrochloric acid and sodium nitrite. This is a somewhat simpler experimental procedure than dealing with gaseous nitrosyl chloride. Significant catalysis by bromide ion and chloride ion has been observed in the nitrosation of alkenes using aqueous nitrous acid.[31] As in nitrosation of many other species (*N*-, *O*-, and *S*-nitrosation), catalysis by the bromide ion is the more substantial, and the nitroso bromo product was isolated. The observations are readily interpreted in terms of the formation of equilibrium concentrations of nitrosyl bromide and nitrosyl chloride in solution, followed by their rate-limiting reactions with the alkene (see scheme (2.38)). For the nitrosation of 2-methylprop-2-enol at $0\,^{\circ}C$ the second-order rate constants for the actual attack by nitrosyl chloride and nitrosyl bromide are 2.7 and $0.471\,mol^{-1}\,s^{-1}$ respectively. This order of reactivity, nitrosyl chloride $>$ nitrosyl bromide, has been confirmed for a large number of different substrates.

$$
\left.
\begin{array}{l}
HNO_2 + H^+ + Br^- \underset{}{\overset{K_{BrNO}}{\rightleftarrows}} BrNO + H_2O \\[4pt]
BrNO + \underset{}{\overset{}{C}=C} \longrightarrow \underset{NO\ \ Br}{C-C}
\end{array}
\right\}
\qquad (2.38)
$$

2.4.2 Nitrosyl carboxylates

Derivatives of propylbenzene when treated with an alkyl nitrite in tri-chloroacetic acid yield β-nitrosotrichloroacetates[32] (equation (2.39)). Similarly a number of acyclic, alicyclic and aryl-substituted alkenes give nitroso carboxylate derivatives when treated with alkyl nitrites or with nitrosonium salts in the presence of the carboxylic acid.[32] In common with many other nitrosations at carbon, the isolated products are blue monomers if the nitroso group is tertiary or colourless dimers if not. The regio-specificity is always that expected from electrophilic nitrosation.

$$
ArCH{=}CHCH_3 \xrightarrow[Cl_3CCO_2H]{RONO} ArCHCH(NO)CH_3
$$
$$
\overset{|}{OCOCCl_3} \qquad (2.39)
$$

It is not known with certainty what the actual nitrosating species is in these cases, but it has been argued[33] that, for example nitrosyl formate is the reagent for the reactions in formic acid, since the reaction takes the same stereochemical course as does the reaction of nitrosyl chloride. In both cases norbornene gives the product of *syn*-addition, a result which is less easy to explain if a nitroso carbocation is formed, which then reacts with formic acid as the nucleophile.

2.4.3 Nitrogen oxides

There is quite a large literature reporting the reactions of a number of nitrogen oxides with alkenes. Some of the reports are conflicting, but certain generalisations can be drawn. Here, these will be restricted to those relating to the formation of nitroso compounds.

In weakly basic solvents such as ethers and esters, dinitrogen tetroxide N_2O_4 appears to react with alkenes by a free-radical mechanism.[34] The major products are usually mixtures of dinitro and nitro-nitrito compounds. Reaction is believed to occur by attack of the nitrogen dioxide radical, the intermediate organic free radical is then captured by another nitrogen dioxide molecule, to give the products by reaction via the nitrogen or oxygen atoms of the second nitrogen dioxide molecule (see scheme (2.40)).

$$(2.40)$$

However, in the absence of weakly basic solvents, nitrosation reactions occur by an ionic mechanism where dinitrogen tetroxide can be regarded as $NO^+NO_3^-$. Typically, the reaction of 2-methylpropene in ethane–propane solution in the temperature range -196 to $-78\,°C$ gives the *anti*-dimeric form of the nitroso nitrate adduct, which is converted to the oxime form upon warming[35] (equation (2.41)). The product structures were established by spectroscopic methods, and also by the identification of the hydrolysis products in some cases,[36] as hydroxy carboxylic acids. An obvious sequence of reactions (scheme (2.42)), involving isomerisation to the oxime, hydrolysis of the nitrate, hydrolysis of the oxime and oxidation of the aldehyde, accounts for the observed products.

$$CH_2{=}C(CH_3)_2 \xrightarrow{N_2O_4} [CH_2(NO)C(ONO_2)(CH_3)_2]_2 \qquad (2.41)$$

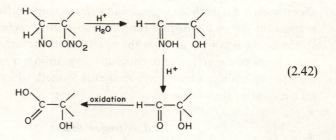

$$(2.42)$$

The change from a free-radical reaction of dinitrogen tetroxide to an ionic mechanism has also been achieved by changing the substituents in the alkene.[37] As hydrophilic substituents are introduced for reaction in aqueous nitric acid, the dinitro and nitro-nitrite products are replaced by the nitroso nitrate products (which undergo further reaction under the experimental conditions). This may be related to a change from a heterogeneous reaction to a homogeneous one as the solubility of the alkene is increased.

Dinitrogen trioxide N_2O_3 also readily reacts with alkenes to give nitroso nitro products, sometimes known as nitrosites. An example is shown in equation (2.43) for the reaction of 2-methylpropene. Reaction can be brought about either with a solution of dinitrogen trioxide in say ether solvent at low temperatures,[39] from aqueous acidic sodium nitrite,[38] or from nitric oxide/nitrogen dioxide mixtures.[39] The product exists as the blue monomer or as a white crystalline dimer. The nitrosites derived from terpenes have proved useful in characterisation, as have the nitroso chloro adducts formed with nitrosyl chloride. The orientation of addition and the nitro rather than nitrito form have been established by a variety of spectroscopic and chemical techniques, although many adducts have been incorrectly formulated in the past. The product derived from 2-methyl-propene has been subjected to a detailed analysis,[40] and a crystal structure of the product from ethylene has been carried out.[41]

$$CH_2{=}C(CH_3)_2 + N_2O_3 \longrightarrow CH_2NO_2C(NO)(CH_3)_2 \quad (2.43)$$

Early mechanistic interpretations[42] accounted for the products in terms of an electrophilic mechanism regarding dinitrogen trioxide to be polarised in the $NO_2^+ NO^-$ sense. There is, however, no evidence in favour of this idea, and it is now believed that the reactions, are free radical in nature,[43,39] probably involving a two-stage process although a direct one-stage reaction involving some kind of four-centre transition state cannot be ruled out.

An alternative interpretation[44] in favour of a nitroso nitrite structure has been suggested on the basis of the observed hydrolysis products of the nitrosites. Hydroxy carboxylic acids are formed (just as from nitroso nitrates), which can readily be interpreted in terms of the nitroso nitrite structure. Interestingly, exactly the same hydrolysis product is obtained from the nitrosite derived from 3-chloro-2-methylpropene, as from the nitroso nitrate from the same alkene and dinitrogen tetroxide. The weight of physical (and other chemical) evidence, however, lies well in favour of the nitro nitroso structures and some other explanation must be sought for the hydrolysis reactions.

In very dilute aqueous solution there is clear kinetic evidence[31] of electrophilic attack by dinitrogen trioxide. The second-order dependence upon $[HNO_2]$ is very diagnostic of the involvement of dinitrogen trioxide and since 2, 3-dimethylbut-2-ene is *ca* 7 times more reactive than 2-methylbut-2-ene, it is clearly an electrophilic process.

2.4.4 *Other nitrosating reagents*

Nitrosyl sulphuric acid reacts with 2-methylbut-2-ene in liquid sulphur dioxide apparently by an electrophilic process.[45] The initial product is relatively unstable and reacts with aqueous alkali to give 3-hydroxy-3-methylbutan-2-one oxime (see scheme (2.44)). A range of trisubstituted alkenes undergo the same reaction[46] to give the hydrogen sulphate oxime expected from an electrophilic nitrosation. From 2, 3-dimethylbutadiene and nitrosyl sulphuric acid a cyclic oxazine is obtained,[47] presumably by a 1, 4-addition as outlined in scheme (2.45).

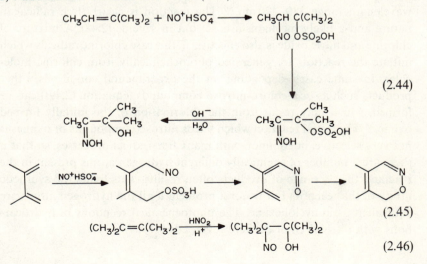

$$CH_3CH{=}C(CH_3)_2 + NO^+HSO_4^- \longrightarrow \underset{\underset{NO\ OSO_2OH}{|\quad\ |}}{CH_3CH\ C(CH_3)_2}$$

(2.44)

(2.45)

$$(CH_3)_2C{=}C(CH_3)_2 \xrightarrow[H^+]{HNO_2} \underset{\underset{NO\ OH}{|\quad\ |}}{(CH_3)_2C{-}C(CH_3)_2}$$

(2.46)

Nitrous acid itself adds to 2, 3-dimethylbut-2-ene and the nitroso alcohol product has been isolated and characterised (see equation (2.46)).[31] The kinetics of the nitrosation of some alkenes by nitrous acid in water show that dinitrogen trioxide or nitrous acidium ion (or nitrosonium ion) can be the reactive species depending on the acidity of the solution.[31] These nitrosating species are readily detected kinetically by either a second- or first-order dependence upon $[HNO_2]$. In the intermediate range it was possible to analyse the data in terms of a mixed first- and second-order reaction (with $[alkene] \gg [HNO_2]$). Substituent effects are again consistent with an electrophilic process for both reagents. When the reagent is nitrous acidium ion (or nitrosonium ion) the kinetic solvent isotope effect k_{D_2O}/k_{H_2O} is 2.2 for the nitrosation of 3-hydroxy-2-methylpropene in sulphuric acid (1.5 M) at 4 °C. This is very similar to the value found for amine nitrosations,[48] and arises from the fact that the concentration of nitrous acidium ion (or nitrosonium ion) is approximately doubled in the deuterated solvent.

2.5 Nitrosation of alkanes

It has been known for some time (since a solution of nitrosyl chloride in heptane was accidentally left in sunlight[49]) that *C*-nitroso compounds or oximes can be obtained from alkanes and nitrosyl chloride in solution or the gas phase, when irradiated with uv–visible light (see scheme (2.47)). High yields can be obtained for a variety of reactants using light in the wavelength range 380–420 nm.[50] The reaction is clearly free radical in nature and a likely mechanism is set out in scheme (2.48). A mixture of chlorine and nitric oxide is also effective; in this case chlorine radicals which initiate the reaction are generated photochemically from chlorine molecules. In some cases, depending on the experimental conditions, other products such as *gem*-chloro-nitroso compounds (equation (2.49)) can be formed, which may result from the chlorination of the initially formed oxime.[51] The initial reaction which yields nitroso compounds or oximes is not very selective, in common with many free-radical processes, so that if there are a number of chemically different hydrogen atoms present in the reactant, then a range of products often results. This limits the synthetic utility of the reaction, but is not a problem if all the hydrogen atoms are equivalent as in cyclohexane. The photochemical reactions of hydrocarbons with nitrosyl chloride were reviewed in 1967.[52]

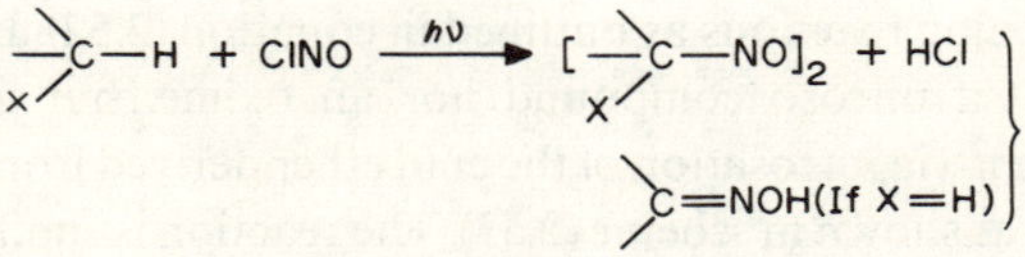

$$\begin{array}{ll} \overset{\diagdown}{\underset{X}{C}}-H + ClNO \xrightarrow{h\upsilon} \left[\overset{\diagdown}{\underset{X}{C}}-NO \right]_2 + HCl \\ \qquad\qquad\qquad\qquad\qquad \overset{\diagdown}{C}=NOH(\text{If } X=H) \end{array} \right\} \qquad (2.47)$$

$$\left. \begin{array}{l} ClNO \xrightarrow{h\upsilon} Cl^{\cdot} + NO \\ Cl^{\cdot} + RH \longrightarrow R^{\cdot} + HCl \\ R^{\cdot} + NO \longrightarrow RNO \end{array} \right\} \qquad (2.48)$$

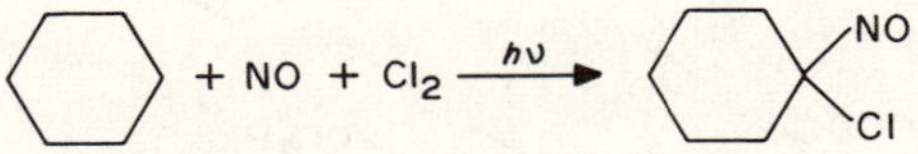

(2.49)

The reaction of cyclohexane and some of its derivatives has considerable industrial application in the synthesis of caprolactam (equation (2.50)), which is the monomer in the production of nylon 6. In this case the photochemical reaction is carried out in acid solution, which catalyses the transformation of the nitroso product into the oxime and also brings about the Beckmann rearrangement to the cyclic amide. The overall reaction can also be achieved by electrophilic nitrosation using nitrosyl sulphuric acid; this is discussed in section 2.2. The alkyl radicals can also be generated by irradiation from a ^{60}Co source,[53] and the oxime (eg of cyclohexanone) is again the isolated product.

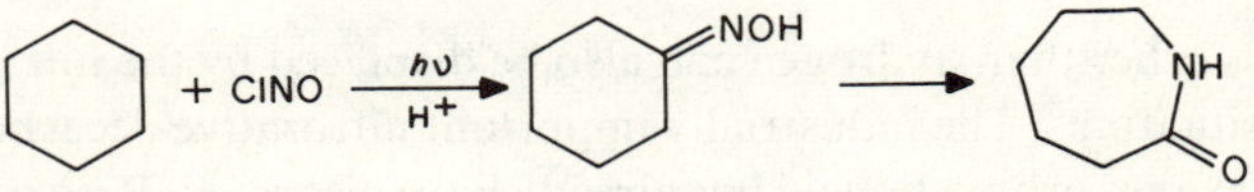

(2.50)

Hydrocarbons possessing an acidic hydrogen atom can readily be nitrosated in basic solution using usually (but not necessarily) alkyl nitrites (scheme (2.51)). Reaction is thought to occur via the carbanion. Fluorene[54] and cyclopentadiene[55] react in this way, but with a sufficiently strong base eg amide salts in liquid ammonia,[56] it is possible to nitrosate much less acidic species such as alkyl benzenes and pyridines.

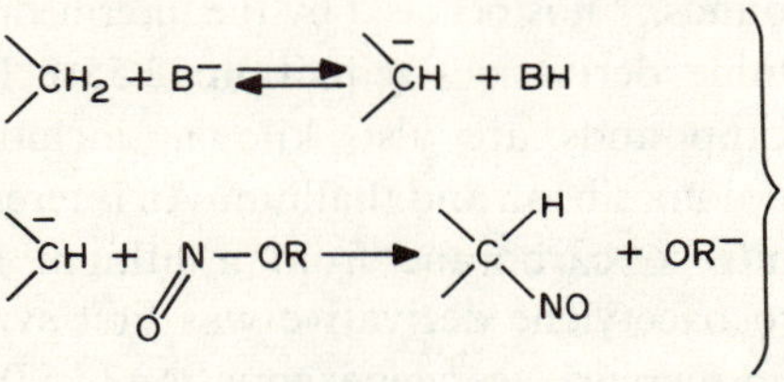

(2.51)

Ketone acetals such as 1,1-diethoxycyclohexane readily undergo nitrosative ring-opening reactions as outlined in equation (2.52). In this case the product is not a nitroso compound nor an oxime, but a cyano ester. Reaction occurs[7] via nitrosation of the enol ether derived from the acetal by loss of ethanol as shown in scheme (2.53). The reaction is similar to the ring-opening reactions which sometimes accompany nitrosation, particularly of enol ethers (see section 2.2).

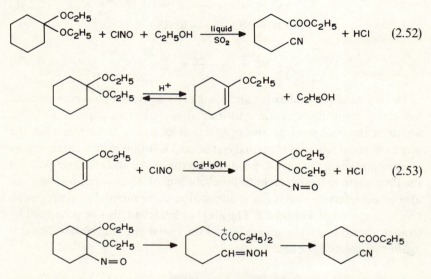

Groups other than hydrogen can also be displaced by the nitroso group in some situations. The industrially important nitrosative decarboxylation of cyclohexane carboxylic acid has already been discussed. Peroxycarboxylic acids react similarly to give nitroso compounds,[57] as shown in equation (2.54).

$$R_2CHCO_3H \xrightarrow{\text{CINO}} R_2CHNO + CO_2 + O_2 \qquad (2.54)$$

A number of organometallic compounds can also be nitrosated. Alkyl (and aryl) Grignard reagents react with, for example nitrosyl chloride to give nitroso compounds,[58] it is believed by the intermediate formation of *N*-nitrosohydroxylamine derivatives (equation (2.55)). Reactions of other organometallic compounds are also known, including derivatives of mercury, tin, aluminium, silicon and thallium. An interesting example is the formation of a nitroso carborane from a lithium derivative (scheme (2.56)).[59] A 1-nitrosoacetylene derivative was first synthesised by nitrosation of an organo-mercury acetylene (equation (2.57)).[60] The product is very unstable and rearranges readily to give a 2-oxoalkanenitrile.

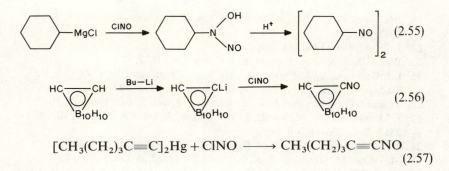

$$[CH_3(CH_2)_3C\equiv C]_2Hg + ClNO \longrightarrow CH_3(CH_2)_3C\equiv CNO \tag{2.57}$$

References

1. S. Dayagi & Y. Degani, in *The Chemistry of the Carbon–Nitrogen Double Bond*, Ed. S. Patai, Interscience, New York, 1970, p. 85.
2. P.A.S. Smith, *Open-chain Nitrogen Compounds*, Benjamin, New York, 1966, Vol. 2, p. 355.
3. J.H. Boyer in *The Chemistry of the Nitro and Nitroso Groups*, Ed. H. Feuer, Interscience, New York 1969, Part 1, p. 215.
4. R.G. Coombes in *Comprehensive Organic Chemistry*, Ed. I.O. Sutherland, Pergamon, Oxford, 1979, Vol. 2, p. 305.
5. C.H. Schmidt, *Angew. Chem.*, 1963, **75**, 169.
6. O. Touster, in *Organic Reactions*, Ed. R. Adams, Wiley, New York, 1953, Vol. 7, p. 327.
7. M.M. Rogic, J. Vitrone & M.D. Swerdloff, *J. Am. Chem. Soc.*, 1977, **99**, 1156.
8. (*a*) K. Singer & P.M. Vamplew, *J. Chem. Soc.*, 1957, 3050.
   (*b*) J.R. Leis, M.E. Pena & D.L.H. Williams, *J. Chem. Soc., Chem. Comm.*, 1987, 45.
9. R.B. Woodward & W. von E. Doering, *J. Am. Chem. Soc.*, 1945, **67**, 860.
10. Y. Ogata, Y. Furuya & M. Ito, *J. Am. Chem. Soc.*, 1963, **85**, 3649; Y. Ogata, Y. Furuya & M. Ito, *Bull. Chem. Soc., Jpn.*, 1964, **37**, 1414.
11. K. Smeykal, W. Pritzkow, G. Mahler, K. Kretschmann & E. Ruehlmann, *J. Prakt. Chem.*, 1965, **30**, 126.
12. I. Donati, G. Sioli & M. Taverna, *Chim. Ind.* (*Milan*), 1968, **50**, 997; L. Giuffre, E. Tempesti, G. Sioli, M. Forneroli & G. Airoldi, *Chem. Ind. Int.* (*Engl. Transl.*), 1971, 1098; *Chem. Ind.* (Milan), 1973, **55**, 258.
13. A.T. Austin, *Sci. Prog.*, 1961, **49**, 619.
14. Y. Furuya, K. Itoho, Y. Kosugi, S. Ishibashi & K. Matsumoto, *Yakugaku Zasshi*, 1981, **101**, 415; *Chem. Abstr.*, 1982, **96**, 19 420; Y. Turuya, K. Itoho, S. Samitani & M. Tomiyama, *Yakugaku Zasshi*, 1981, **101**, 999; *Chem. Abstr.*, 1982, **96**, 103 382.
15. V. Meyer, *Ber. Dtsch. Chem. Ges.*, 1873, **6**, 1492.
16. A.T. Nielsen in *The Chemistry of the Nitro and Nitroso Groups*, Ed. H. Feuer, Interscience, New York, 1969, Part 1, p. 394.
17. E.H. Bartlett & M.D. Johnson, *J. Chem. Soc. A*, 1970, 523.

18. P.S. Khokhlov, B.A. Kashemirov & Y.A. Strepikheev, *Zh. Obshch. Khim.*, 1982, **52**, 2800.
19. J. Thiele, *Ber.*, 1900, **33**, 669.
20. W.A. Tilden & W.A. Stenstone, *J. Chem. Soc.*, 1877, 554.
21. L.J. Beckham, W.A. Fessler & M.A. Kise, *Chem. Rev.*, 1951, **48**, 319; P.P. Kadzyauskas & N.S. Zefirov, *Russ. Chem. Rev.*, 1968, **37**, 543.
22. T. Beier, H.G. Hauthal & W. Pritzkow, *J. Prakt. Chem.*, 1964, **26**, 304.
23. J. Meinwald, Y.C. Meinwald & T.N. Baker, *J. Am. Chem. Soc.*, 1963, **85**, 2513; *ibid.*, 1964, **86**, 4074.
24. Y.L. Chow, K.S. Pillay & H. Richard, *Canad. J. Chem.*, 1979, **57**, 2923.
25. M. Ohno, M. Okamoto & K. Nukada, *Tetrahedron Lett.*, 1965, 4047.
26. See P.B.D. de la Mare & R. Bolton, *Electrophilic Additions to Unsaturated Systems*, Elsevier, Amsterdam, 1966.
27. M.M. Rogic, T.R. Demmin, R. Fuhrmann & F.W. Koff, *J. Am. Chem. Soc.*, 1975, **97**, 3241.
28. K.A. Oglobin & D.M. Kunuwskaya, *Zh. Org. Khim.*, 1968, **4**, 897.
29. J.R. Mahajan, G.A.L. Ferreira & H.C. Araujo, *J. Chem. Soc., Chem. Commun.*, 1972, 1078.
30. R.H. Reitsema, *J. Org. Chem.*, 1958, **23**, 2038.
31. J.R. Park & D.L.H. Williams, *J. Chem. Soc., Perkin Trans. 2*, 1972, 2158.
32. A. Bruckner & F. Ruff, *Acta Chim. Acad. Sci. Hung.*, 1963, **32**, 129; *Chem. Abstr.*, 1963, **59**, 13854.
33. H.C. Hamann & D. Swern, *J. Am. Chem. Soc.*, 1968, **90**, 6481; P. Gray & A.D. Yolfe, *Chem. Rev.*, 1955, **55**, 1069.
34. H. Schechter, *Rec. Chem. Progr.*, 1964, **25**, 25.
35. L. Parts & J.T. Miller, *J. Phys. Chem.*, 1969, **73**, 3088.
36. E.F. Schoenbrunn & J.H. Gardner, *J. Am. Chem. Soc.*, 1960, **82**, 4905; B.F. Ustavschikov, V.A. Podgornova & M.I. Faberov, *Dokl. Akad. Nauk SSSR*, 1966, **168**, 1335; B.F. Ustavschikov, V.A. Podgornova, N.V. Dormidontova & M.I. Faberov, *ibid.*, 1964, **157**, 143.
37. J.B. Wilkes & R.G. Wall, *J. Org. Chem.*, 1980, **45**, 247.
38. C. Belzecki, *Bull. Acad. Polon. Sci., Ser. Sci. Chim.*, 1968, **11**, 129; *Chem. Abstr.*, 1963, **59**, 15199.
39. L. Jonkman, H. Muller & J. Kommandeur, *J. Am. Chem. Soc.*, 1971, **93**, 5833.
40. J. Pfab, *J. Chem. Soc., Chem. Commun.*, 1977, 766.
41. F.P. Boer & J.W. Turley, *J. Am. Chem. Soc.*, 1969, **91**, 1371.
42. C.K. Ingold & E.H. Ingold, *Nature*, 1947, **159**, 743.
43. A. Michael & G.H. Carlson, *J. Org. Chem.*, 1940, **5**, 1.
44. J.R. Park & D.L.H. Williams, *J. Chem. Soc., Perkin Trans. 2*, 1976, 828.
45. D.G. Boller & G.H. Whitfield, *J. Chem. Soc.*, 1964, 2773.
46. W. Kisan & W. Pritzkow, *J. Prakt. Chem.*, 1978, **320**, 59.
47. D. Klamann, M. Fligge, P. Weyerstahl & J. Kratzer, *Chem. Ber.*, 1966, **99**, 556.
48. B.C. Challis, L.F. Larkworthy & J.H. Ridd, *J. Chem. Soc.*, 1962, 5203.
49. E.V. Lynn, *J. Am. Chem. Soc.*, 1919, **41**, 368.
50. E. Müller & H. Metzger, *Chem. Ber.*, 1957, **90**, 1179.

51. O. Piloty, *Ber.*, 1898, **31**, 452.
52. M. Pape, *Fortschr. Chem. Forsch.*, 1967, **7**, 559.
53. P.R. Hills & R.A. Johnson, *Int. J. Appl. Radiation and Isotopes 12*, 1961, **80**, Nos. 3/4.
54. C.F. Koelsch, *J. Org. Chem.*, 1961, **26**, 1291.
55. J. Thiele, *Ber.*, 1900, **33**, 666.
56. S.E. Forman, *J. Org. Chem.*, 1964, **29**, 3323.
57. M.M. Labes, *J. Org. Chem.*, 1959, **24**, 295.
58. E. Müller & H. Metzger, *Chem. Ber.*, 1956, **89**, 396.
59. J.M. Kauffman, J. Green, M.S. Cohen, M.M. Fein & E.L. Cottrill, *J. Am. Chem. Soc.*, 1964, **86**, 4210.
60. E. Robson & J.M. Tedder, *Proc. Chem. Soc.*, 1963, 13.

3

Aromatic C-nitrosation

Aromatic nitration is a very well-known reaction which is much used synthetically and which has also been widely studied mechanistically. The topic is comprehensively covered in a manuscript by Schofield.[1] By contrast aromatic *C*-nitrosation is much less known and has been studied mechanistically in a relatively few cases. The reason for this is that substantial activation by electron-releasing groups is necessary to bring about nitrosation at a convenient rate. The implication is that the nitrosating agents nitrosonium ion, nitrous acidium ion and XNO are much less powerful electrophiles than is the nitronium ion NO_2^+, or that they are generated in much lower concentrations. Alternatively, the mechanisms of the two reactions may be different. It transpires that each of these factors contributes to the lower reactivity in nitrosation. It has been estimated[2] that the nitrosonium ion is less reactive than is the nitronium ion by a factor of *ca* 10^{14}.

So far as the preparation of aromatic nitroso compounds is concerned, routes involving the direct substitution of a hydrogen atom are mostly limited to reactants containing hydroxy, alkoxy, primary and secondary amine groups. Other nitroso compounds have been synthesised by the direct substitution of groups other than the hydrogen atom ie by *ipso* substitution. Examples include the reaction of some carbonyl compounds and also of some organometallic species. Most of the reagents discussed in chapter 1 can be used in aromatic nitrosation, contrasting with the situation in nitration where the vast majority of reactions have involved the nitronium ion. Because of the structural limitations in nitrosation, much less is known mechanistically compared with nitration. There are very few results on isomer ratios of products, relatively few rate measurements and generally little is known about structure–reactivity relationships. Nevertheless, some mechanistic features have been established, notably for the reactions of phenols, where significant differences occur between nitration and nitrosation. The two reactions are, however, closely linked in the area of nitration reactions which are catalysed by nitrous acid, often present in nitric acid in small concentrations. This topic has been reinvestigated and a new reaction pathway has emerged for aromatic nitration, which is discussed in detail in section 3.3.

3.1 Products

The most familiar reactions are those of phenols and naphthols. Both are readily nitrosated in dilute acid solution by nitrous acid at room temperature to give *C*-nitroso compounds. In solution these species exist primarily as the tautomeric benzo (or naphtho) quinone monooximes. Phenol itself gives primarily 4-nitrosophenol,[3] with about 10% of the 2-isomer at 40 °C, as outlined in equation (3.1). Similarly 2-naphthol gives quantitatively 1-nitroso-2-naphthol[4] (equation (3.2)). One feature of aromatic nitrosation reactions is that the isolated products are sometimes not the nitroso compounds but the nitro compounds. This occurs particularly when excess nitrous acid is used and it is thought that the first-formed nitroso compound is oxidised to the nitro compound by nitrous acid.

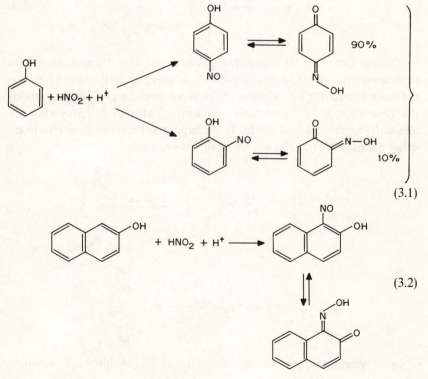

(3.1)

(3.2)

Phenyl ethers are also readily nitrosated, but these reactions are often accompanied by dealkylation, so that, for example, for both anisole (equation (3.3)) and diphenyl ether,[2] the main isolated product of nitrosation is 4-nitrosophenol. Resorcinol diethyl ether yields a mixture of products (equation (3.4)), the nitroso dialkyl ether and the phenol ether derived from dealkylation of one of the ether groups.[5] The mechanism of dealkylation is not known with certainty, but may arise from hydrolysis of the nitroso ether first formed, although this does not appear to take place in the halogenation of phenyl ethers.

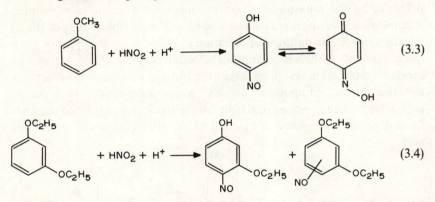

$$+ \; HNO_2 + H^+ \longrightarrow \qquad \rightleftharpoons \qquad \qquad (3.3)$$

$$+ \; HNO_2 + H^+ \longrightarrow \qquad + \qquad \qquad (3.4)$$

Phenols can also be nitrosated photochemically. Phenol itself with aqueous nitrite ion in the presence of uv light gives again 4-nitrosophenol as the main product[6] (see equation (3.5)). Here reaction probably occurs by the generation (and subsequent reaction) of nitroso hydroxy radicals. Similarly naphthols and anthrols undergo photochemical nitrosation,[7] using nitrosamines as the source of nitroso radicals.

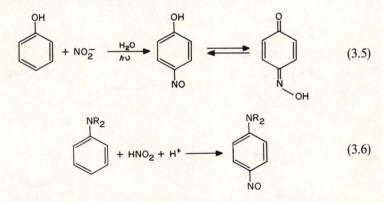

$$+ \; NO_2^- \; \xrightarrow[h\upsilon]{H_2O} \qquad \rightleftharpoons \qquad \qquad (3.5)$$

$$+ \; HNO_2 + H^+ \longrightarrow \qquad \qquad (3.6)$$

Tertiary aromatic amines react with nitrous acid in dilute acid solution, as do phenols, to give mainly the 4-nitroso product (equation (3.6)). In some cases there is also evidence of some 2-substitution,[8] but the 2-nitrosoamine is oxidised to the 2-nitroamine, so that the isolated product is the nitro and not the nitroso compound. It appears that 2-nitrosoamines are not particularly stable compounds and are readily oxidised. Dealkylation can also occur with tertiary amine nitrosation, particularly (as for phenyl ethers) in the presence of excess of nitrous acid. Under these conditions there is, for example, substantial conversion of N, N-dimethylaniline into N-methyl-4-nitro-N-nitrosoaniline,[9] probably involving the sequence of steps, C-nitrosation, oxidation, dealkylation and N-nitrosation outlined in scheme (3.7). The reaction rate is much reduced in the presence of bulky N-alkyl groups and also by 2-substituents,[10] no doubt as a result of a steric effect.

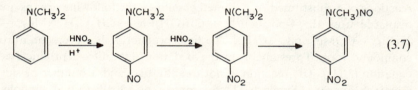

$$(3.7)$$

Secondary aromatic amines are readily nitrosated at the nitrogen atom, and the N-nitrosamine can then rearrange intramolecularly in acid solution to form the 4-nitroso derivative (see scheme (3.8)). This is the Fischer–Hepp rearrangement, and is discussed in detail in section 5.1.

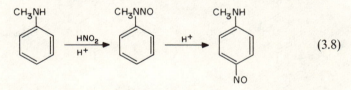

$$(3.8)$$

Benzene and alkylbenzenes react slowly with nitrous acid in concentrated sulphuric acid.[2] Nitrosoarenes are probably the initial products but these react further and diazonium salts are formed;[11] this is a known reaction of nitrous acid with C-nitroso compounds.

Nitroso phenols can also be generated from benzene derivatives using hydroxylamine, hydrogen peroxide and either copper(II) salts or sodium pentacyanoammineferroate (equation (3.9)). This is the Baudisch reaction[12] which almost certainly involves free-radical intermediates.

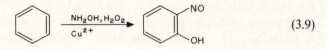

$$(3.9)$$

Aromatic nitroso compounds can also be prepared by the substitution by NO, of groups other than the hydrogen atom ie by *ipso* substitution. The best-known example is the nitrosative decarboxylation of benzoic acid derivatives.[13] This reaction, which is quantitative for the 3, 5-dibromo-4-hydroxy compound (equation (3.10)), occurs readily at room temperature with the immediate evolution of carbon dioxide.

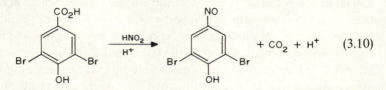

$$+ CO_2 + H^+ \qquad (3.10)$$

Displacement of other groups can also occur, for example, in the reactions of 4-substituted N, N-dimethylanilines, the following groups are displaced from the 4-position[14]: $—CH(OH)Ph$, $—CH_2C_6H_4NMe_2$, $—N_2Ph$, $—COMe$, and $—CHO$. The final product is, however, the nitro compound, formed presumably by the oxidation of the nitroso product (see equation (3.11)). The reaction is not quantitative, and a number of side reactions occur. Similarly,[14] a range of 3-substituted indoles (equation (3.12)) and 3-substituted indolizines react with nitrous acid by *ipso* attack at the 3-position to give a mixture of the 3-nitroso and 3-nitro derivatives as the major products.

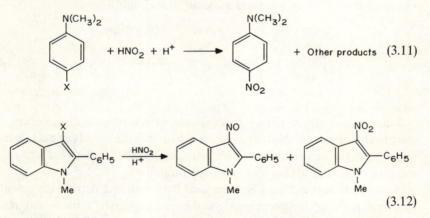

$$(3.12)$$

A number of organometallic compounds react with a range of nitrosating agents (nitrosyl halides, dinitrogen trioxide, dinitrogen tetroxide, and alkyl nitrites) to form nitrosoarenes, by the displacement of the metallic group. Examples include the reactions of aryl mercury halides, aryl Grignard reagents, tin, silicon, and thallium compounds.[15] The reaction is typified in equation (3.13) for the Grignard reagent.

$$\text{(3.13)}$$

None of the methods described involving *ipso* substitution is a particularly suitable synthetic route to aryl nitroso compounds generally, since the yields are often not high and other products (notably the nitro compounds) are often formed concurrently. The direct substitution of the hydrogen by the nitroso group also has limited synthetic utility, again because of oxidation problems and also the limitation to amines and phenols. Better synthetic methods are available, particularly those based on the reduction of the nitro compounds and the oxidation of amines, and these have been discussed elsewhere.[16]

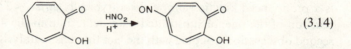

$$\text{(3.14)}$$

Non-benzenoid aromatic systems may also be *C*-nitrosated.[17] These reactions have some synthetic applications, but have not been much studied. Fulvenes, tropolones, and azulenes are readily nitrosated; equation (3.14) shows the formation of 5-nitrosotropolone.

3.2 Reaction mechanisms

Most of the detailed reaction mechanism studies of aromatic *C*-nitrosation have centred upon the reactions of phenols, naphthols and aryl ethers. An important contribution to the understanding of these reactions comes from a series of papers by Challis and coworkers in the 1970s. The experimental facts can be accommodated by a mechanism involving the reversible formation of an intermediate, followed by a proton transfer to the medium. This is the familiar $A–S_E2$ mechanism, well known in other aromatic electrophilic substitution reactions, including nitration and halogenation. For a phenol or a naphthol the intermediate is thought to be a neutral dienone species as shown in scheme (3.15) for the reaction of phenol itself with a general nitrosating species XNO. The dienone intermediate has not been detected spectroscopically (contrasting with the situation in bromination of phenols) but may be too reactive to build up in a significant concentration.

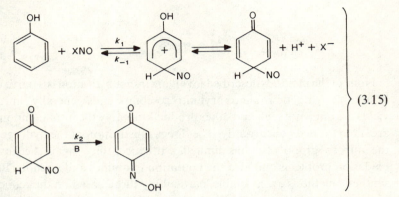

$$(3.15)$$

Two unusual features occur in the nitrosation of phenols: (*a*) there is no overall acid catalysis, at least in the acid range pH 1.0–4.5,[18,19] and (*b*) there is a pronounced kinetic isotope effect (typically $k_H/k_D \sim 3.5$),[18,20] for the reaction of the phenol substituted with deuterium in the 4-position. The second observation, coupled with the fact that base catalysis (eg by acetate ion) also occurs, indicates that under these conditions, the final proton transfer to the medium (indicated by B in scheme (3.15)) from the aromatic ring is at least, in part, rate-limiting. This is rather uncommon in aromatic electrophilic substitution particularly in nitration and halogenation, where usually the rate-limiting step is the attack of the electrophile. Rate-limiting proton loss is known in these and other reactions, but usually when rather special structural conditions (often steric) apply. The expected general rate equation from the mechanism outlined in scheme (3.15) is given in equation (3.16) for reaction with nitrous acid in acid solution, in the presence of X^-. When the pH is greater than *ca* 2 it is necessary to make allowance for the ionisation of nitrous acid to nitrite ion. This has not been included here, for simplicity.

$$\text{Rate} = \frac{k_1 k_2 [C_6H_5OH][H^+][HNO_2][B][X^-]}{k_{-1}[H^+][X^-] + k_2[B]} \qquad (3.16)$$

In the absence of added X^- (ie when $X^- = H_2O$), it is easy to see that, when the limit $k_{-1}[H^+] \gg k_2[B]$ applies, a primary kinetic isotope effect and general base catalysis is predicted, together with the absence of acid catalysis. Similarly when $k_{-1}[H^+] \ll k_2[B]$ ie at lower acid concentration, the reaction should be acid-catalysed with no kinetic isotope effect nor base catalysis. The former situation is the more common, but both of these limiting conditions have been achieved, both for phenol and also for 2-

naphthol.[18,19] Where base catalysis occurs, the Brønsted equation is followed with a β value for phenol of 0.37.

Further confirmation of this mechanism comes from the effect of added nucleophiles (X^-) chloride ion, bromide ion and thiocyanate ion. When $k_{-1}[H^+][X^-] \gg k_2[B]$ there is no nucleophilic catalysis as expected when the final proton transfer is rate-limiting. However, at low acidity and low $[X^-]$ there is nucleophilic catalysis,[21] in the expected order, chloride ion < bromide ion < thiocyanate ion. The observed rate constant is, however, not linearly dependent upon $[X^-]$, which suggests that the limiting condition $k_{-1}[H^+][X^-] \ll k_2[B]$ is not fully achieved. This occurs also in some N- (and other) nitrosations.

As expected the second-order rate constants for attack by nitrosyl chloride, nitrosyl bromide and nitrosyl thiocyanate lie in the order nitrosyl chloride > nitrosyl bromide > nitrosyl thiocyanate (for reaction with 2-naphthol[21]) but the values (5.5×10^7, 3.3×10^6, and $1.4 \times 10^4 \, l \, mol^{-1} \, s^{-1}$ respectively) are much below those for the corresponding reactions of aromatic amines (such as aniline) where both nitrosyl chloride and nitrosyl bromide react at the diffusion limit.

In the absence of added catalysts reaction at low acid concentrations appears to involve nitrous acidium ion (or nitrosonium ion), there being no evidence of reaction via dinitrogen trioxide. The third-order rate constant for the reaction of 2-naphthol is $4700 \, l^2 \, mol^{-2} \, s^{-1}$ at $25 \, °C$ and so is within the range (2000–$6000 \, l^2 \, mol^{-2} \, s^{-1}$) where reaction is believed to be diffusion-controlled.

A study has been made of the reactions of five 4-substituted phenols.[22] In each case the 2-nitro product was formed, by rate-limiting nitrosation and subsequent oxidation. The rate constants are increased by electron-donating substituents and there is a good Hammett correlation with $\rho = -6.2$. Below $0.5 \, M \, H^+$ the reaction is not acid-catalysed, and there is a kinetic isotope effect on the loss of the ring proton. All these facts suggest that the mechanism proposed for phenol (scheme (3.15)) applies also here.

At acidities greater than $\sim 0.5 \, M$, acid catalysis is a general feature of phenol nitrosation. This is believed to arise by the onset of another pathway for the decomposition of the dienone intermediate which is acid-catalysed, and which competes with the spontaneous decomposition (see equation (3.17)). This additional pathway probably involves protonation of the nitroso oxygen atom. The rate–acid concentration profile has a maximum at ~ 6–$7 \, M$ perchloric acid and at higher acid concentrations the rate constant decreases, probably because of the decreasing water concentration in this region, which acts as the base for the final proton removal.

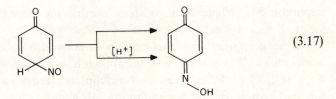

(3.17)

Phenyl ethers react similarly to phenols, except that there is no acid-independent pathway at moderately low acidities. Rate measurements have been reported for the reaction of anisole[18,2,23] and for diphenyl ether.[2] All the data fit the expected A–S_E2 mechanism (see scheme (3.18)) with an intermediate σ complex; a dienone intermediate is not now possible. Again the reactions are characterised by aromatic ring proton kinetic isotope effects, so that proton loss is rate-limiting. A consequence is that no direct deductions can be made about the nature of the nitrosating agent. The product is always the 4-nitroso phenol (or its tautomer), believed to arise from hydrolysis of the 4-nitroso phenyl ether.

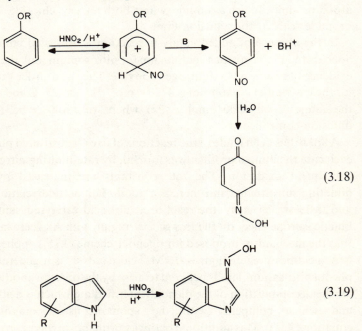

(3.18)

(3.19)

C-Nitrosation has also been studied mechanistically in the heteroaromatic indole systems.[24] A number of 1-, 2-, and 5-substituted indoles have been examined. The product in each case is the tautomeric oxime form of the 3-nitroso product (see equation (3.19)). The same mechanism applies here as for the reactions of phenols and phenyl ethers, but now the rate-limiting step depends on the reactivity of the indole. For the more basic compounds (with p$K_a \geqslant -3.5$) formation of the Wheland intermediate is rate-limiting. The nitrosating agents nitrous acidium ion (or nitrosonium ion), dinitrogen trioxide, and nitrosyl chloride have each been identified kinetically, and the values of the rate constants indicate that the process is diffusion-controlled. As an example, for the reaction of 1,2-dimethylindole with nitrosyl chloride, the rate constant is $5.4 \times 10^8 \, \mathrm{l \, mol^{-1} \, s^{-1}}$ at $0\,°\mathrm{C}$, compared with a value of $2.2 \times 10^9 \, \mathrm{l \, mol^{-1} \, s^{-1}}$ for aniline at $25\,°\mathrm{C}$. In the case of the less basic indoles (p$K_a \leqslant -6.0$) the second stage, ie proton loss, is rate-limiting, in common with many other aromatic nitrosations. As for the reaction of phenols, there is an acid-catalysed pathway for the final step in the sequence. The very high reactivity of indoles compared with phenols and phenyl ethers, suggests that reaction may, in fact, occur initially at the nitrogen atom, and that this is followed by an internal nitroso group rearrangement to the 3-position. Such rearrangements are now reasonably well recognised in some other nitrosations.

(3.20)

Kinetic measurements have also been reported for the nitrosation of azulene and of 1-nitroazulene,[25] whose basicities are comparable with those of the indoles reported earlier. For azulene (see equation (3.20)) there is chloride ion catalysis and no primary ring hydrogen kinetic isotope effect, suggesting that the formation of an intermediate is rate-limiting, again probably taking place at the encounter limit. The less reactive 1-nitroazulene, however, is not subject to chloride ion catalysis in nitrosation and there is a primary kinetic isotope effect, strongly suggesting that the final proton loss is rate-limiting. The result with azulene in particular suggests that all reactive aromatic systems with p$K_a \geqslant -3.5$ react with nitrous acidium ion (or nitrosonium ion) and nitrosyl chloride at the encounter rate, and that such a reaction can indeed take place at carbon. This result suggests that the more basic indoles might also undergo a direct nitrosation at carbon, and that it is not necessary to invoke $N \rightarrow C$ internal rearrangement of the nitroso group. The nitrosation of benzene and some alkyl benzenes has also been examined kinetically. Here the conditions are necessarily more forcing because of the low reactivity of the substrate; typical conditions are $\sim 10\,\mathrm{M}$ perchloric acid at $53\,°\mathrm{C}$. Under these conditions decomposition of nitrous acid (or of nitrosonium ion) is a serious problem, so that the rate constants are much less reliable than are those obtained at lower temperatures. Early measurements[2] were taken to support the general A–S_E2 mechanisms with rate-limiting proton loss. However, more recent experiments[26] particularly with toluene, *o*-xylene and mesitylene have cast some doubt upon this interpretation. Both nitration and nitrosation appear to take place concurrently, the former via the conventional nitronium ion mechanism, but the latter is believed not to be an electrophilic process, but results from the decomposition products of nitrous acid (various nitrogen oxides). The detailed mechanism has not been established. The decomposition of nitrous acid makes it a very difficult problem. There is another report[27] of the nitrosation of aromatic hydrocarbons in trifluoroacetic acid using sodium nitrite. Here the products are exclusively nitroarenes, formed in high yield. The product analysis evidence argues against a nitrosation/oxidation mechanism but rather suggests that the nitronium ion is involved, formed by the sequence $HNO_2 \rightarrow N_2O_3 \rightarrow NO_2(+ NO) \rightarrow N_2O_4 \rightarrow NO_2^+$.

The nitrosation of benzene has been studied from a theoretical point of view.[28] Electron transfer takes place from a benzene frontier orbital to a $\pi*$ orbital in a nitrosonium ion to give two interconverting π complexes of different symmetry. Such complexes have been detected in reactions in solution and in the gas phase. The spectra of the product of nitrosation of benzene have been obtained, both in the gas phase (from benzene and methyl nitrite) and in solution (from benzene and nitrosonium ion).[29] The two spectra are very similar and are thought to represent $C_6H_6NO^+$ as a π complex of charge-transfer type, where the charge is located mainly in the aromatic ring.

There have been no detailed mechanistic studies of the *C*-nitrosation of *N, N*-dialkylaniline derivatives even though the reaction is well known preparatively, and indeed gives the 4-nitroso compound almost quantitatively under certain conditions.[9] The 2-nitroso amine is not known, but a little 2-nitro-*N, N*-dimethylaniline is formed from *N, N*-dimethylaniline. When the 4-position is blocked as in 4-methyl-*N, N*-dimethylaniline, the 2-nitro compound is formed (see equation (3.21)).[30] Since nitrosation of secondary amines occurs readily at nitrogen, and the rearrangement to give the 4-nitroso compounds (the Fischer–Hepp rearrangement) occurs intramolecularly, it seems likely that the same situation applies to the *C*-nitrosation of tertiary aromatic amines, ie that *N*-nitrosation first occurs, followed by an intramolecular nitroso group rearrangement (see scheme (3.22)). This has never been demonstrated, and indeed, it is not easy to see how this could be done experimentally since the salts of *N*-nitroso tertiary aromatic amines are not stable compounds. It is possible that this sort of mechanism applies more generally eg in the nitrosation of phenols, naphthols, and phenyl ethers, where *O*-nitrosation could occur (well known with alcohols) followed by an intramolecular rearrangement, although again it is difficult to see how this could be proved experimentally.

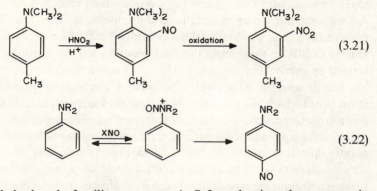

$$(3.21)$$

$$(3.22)$$

It is likely that the familiar two-stage A–S_E2 mechanism also operates in those cases where a group other than a hydrogen atom is displaced. This is written out in scheme (3.23) for the reaction of 3,5-dibromo-4-hydroxybenzoic acid. Kinetic measurements,[20] including the variation of rate constants with pH, are consistent with such a mechanism occurring via the dianionic species. Similar mechanisms have been advocated for bromodecarboxylation and bromodesulphonation reactions. 4-Methoxybenzoic acid is ∼ 300 times less reactive than is 4-hydroxybenzoic acid, which implies that either reactivity is conferred by the —O⁻ group or that the transition state leading to the dienone intermediate is of lower energy than that leading to the Wheland intermediate for the reaction of the ether.

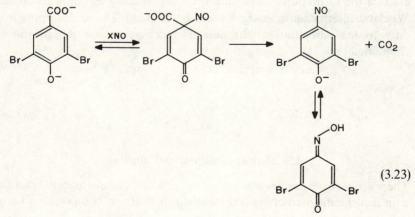

$$(3.23)$$

A different explanation has been put forward[14] to account for the mechanism of *ipso* nitrosation in 3-substituted indoles, N,N-dimethylanilines, and 1-, and 3-substituted indolizines. This involves initially a one-electron transfer from XNO (the carrier of the nitrosonium ion) to give a radical cation and nitric oxide which react further by a free-radical reaction at an aromatic carbon atom to give the Wheland intermediate. The product is then formed by expulsion of Y as Y^+ (see scheme (3.24)). In these cases the final products are the nitro compounds formed by subsequent oxidation. This mechanism is similar to one which has been suggested on several occasions for electrophilic aromatic nitration,[31] but which has not attracted widespread support. In the case of the *ipso* nitrosation reactions in the indoles, dimethylanilines and indolizines,[14] the evidence in favour of radical cation intermediates is wholly derived from a rationalisation of the by-products in these terms, and also from a consideration of the appropriate standard electrode potentials. There is, at the present, no esr spectral, or other evidence in favour of such a mechanism, and the results are equally well accommodated in terms of a conventional two-electron transfer electrophilic nitrosation mechanism. However, it is well known that nitrosonium salts (in eg acetonitrile) and solutions of nitrous acid at high acidity, where it is known that substantial concentrations of nitrosonium ion exist, can also effect one-electron oxidation reactions. These have been discussed in more detail in section 1.5. In addition it is now known that the first step of nitrous-acid-catalysed nitration involves a one-electron transfer to nitrosonium ion to give the radical cation. These reactions occur at high acid concentration (and are discussed fully in the next section). It is possible that the first stage of *any* aromatic nitrosation, where significant nitrosonium ion concentrations exist, is the formation of the radical cation which then collapses to the Wheland intermediate, as shown in scheme (3.24). This could apply not only to *ipso* substitution but also to reactions where the proton is eliminated (Y = H).

$$\text{ArY} + \text{XNO} \longrightarrow \text{ArY}^{\bullet+} + \text{NO} + \text{X}^-$$
$$\downarrow$$
$$\text{ArNO} + \text{Y}^+ \longleftarrow \text{Ar}^+\overset{\text{Y}}{\underset{\text{NO}}{\diagdown}} \qquad (3.24)$$

3.3 Nitrous-acid-catalysed nitration

There are many examples where the final product of a nitrosation reaction is in fact the nitro compound rather than the nitroso compound. This is particularly true if nitrous acid is present in excess (which can act as an oxidising agent), for some *ipso* nitrosation reactions, and also when nitrosation takes place *ortho* to an amino group in an aromatic system. Such reactions have synthetic potential, since the final nitro product in some cases is difficult to synthesise by a conventional nitronium ion nitration. A number of these reactions have already been discussed and there seems to be no doubt that reaction involves initial nitrosation followed by oxidation. However, in addition to these examples, it has been known for a long time[32] that nitrous acid (or some species derived from it) plays an important part in the nitration of some reactive aromatic systems, such as phenols and amines. Experimentally it was found that these nitrations were catalysed to a significant extent by the presence of small quantities of nitrous acid, either added as sodium nitrite or present as a decomposition product in nitric acid. Further the isomer ratios of products were often different under these conditions from those formed in the 'normal' nitration reaction. This catalysed pathway can readily be eliminated by removing all traces of nitrous acid by the addition of sulphamic acid, hydrazine, urea or any other species known to react rapidly and irreversibly with nitrous acid. The catalysed reaction can also, in some situations, be made the dominant pathway by the addition of sufficient nitrous acid.

There is also an anticatalytic effect of nitrous acid in aromatic nitration, which occurs with a much wider range of substrate. The effect is quite small and appears to be the result of effectively decreasing the concentration of the nitronium ion responsible for nitration.

Until recently it was generally believed that all nitrous-acid-catalysed nitration reactions occurred by an initial nitrosation followed by oxidation of the nitroso group to the nitro group by nitric acid (or some species derived from it). Nitrous acid would be expected to be formed in this second stage (by reduction of nitric acid), so it is easy to understand that only small catalytic amounts of nitrous acid would be necessary. Such small amounts, formed by decomposition, are generally present in nitric acid, unless steps are taken to remove them. Another feature resulting from this sequence is autocatalysis, which has been observed in the nitration of anthracene.[33]

Kinetic measurements[34] on the nitration of 4-nitrophenol by nitric acid in the presence of nitrous acid, were taken to support this mechanism. Reaction was first order in both [4-nitrophenol] and [HNO_2] in acetic acid solvent, and the variation with acid suggested that the reactive species was the 4-nitrophenolate ion (see scheme (3.25)).

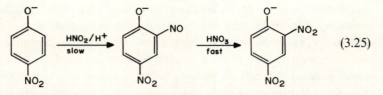

$$ (3.25) $$

However, more recent observations cannot be explained by this mechanism. In a study[35] of the nitration of *N, N*-dimethylaniline in the presence of nitrous acid, the reaction was found to be considerably faster than the *C*-nitrosation reaction performed in the absence of nitric acid. Further, in addition to the expected nitration product (4-nitro-*N, N*-dimethylaniline) significant amounts of *N, N, N', N'*-tetramethylbenzidine were formed, in yields which increased with the acidity. These results cannot be explained in terms of the outline mechanism given in scheme (3.25) and an alternative was suggested (given in scheme (3.26)) involving a one-electron transfer. The first stage is thought to be the formation of a radical cation–radical pair (with nitric oxide) which can either dissociate in nitric acid to give the free cation radical which dimerises to give the benzidine product, or which can exchange with the nitronium ion to give another radical cation–radical pair, this time with nitrogen dioxide, from which the nitro product derives.

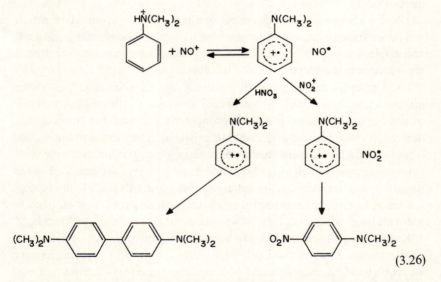

$$ (3.26) $$

Support for this mechanism comes from the observation of CIDNP effects when the reactions are carried out with ^{15}N-labelled nitric-acid,[36] and also by the observation of esr spectra from the reaction solutions,[37] which are consistent with radical cation structures eg $4\text{-}CH_3C_6H_4N(CH_3)_2^{+\cdot}$.

It now appears that this mechanism also accounts for the experimental observations in the nitrous-acid-catalysed nitration of other substrates eg phenols, phenyl ethers and also reactive alkyl benzenes such as mesitylene. The product yields and the *ortho:para* ratio from experiments with phenol[38] using nitric acid alone, nitrous acid alone, and mixtures of nitric and nitrous acids cannot be explained by nitrosation/oxidation, but fit well into scheme (3.26). Similarly the kinetics of 4-nitrophenol nitration in aqueous media also support this new mechanism.[39] At low nitrous acid concentrations the reaction is first-order with respect to both the [substrate] and $[HNO_2]$ (a result also found earlier for nitration in acetic acid[34]) – but as the concentration of nitrous acid is increased the reaction becomes zero-order in $[HNO_2]$ whilst remaining first-order in [substrate]. In scheme (3.26), these two limiting cases represent, respectively, rate-limiting formation of the first radical cation–radical pair (with nitric oxide), and rate-limiting attack by nitrogen dioxide to give the final nitro product. Similarly a detailed kinetic study of the nitrous-acid-catalysed nitration of phenol has been interpreted[40] by the same mechanism; a small component of the reaction takes place by nitrosation and then oxidation of the nitroso compound. Rate data on the nitrous-acid-catalysed nitration of phenyl ethers[23,41] have also been interpreted by the one-electron transfer mechanism, which in some cases is accompanied by a direct two-electron transfer nitrosation followed by oxidation. Mesitylene also undergoes nitrous acid catalysed nitration.[42] Again CIDNP effects are observed which are totally absent when the pathway involving nitrous acid is eliminated by the addition of azide ions. The actual nuclear polarisation effect is thought to arise from the partitioning of the radical ion–radical pair $ArH^{+\cdot}NO_2^{\cdot}$ between dissociation to separate radicals and combination to give the Wheland intermediate and thence the nitro product. The formation of some side products (in low yields) can also be rationalised in terms of other reactions of the radical cation intermediate.

It now seems likely that a range of substrates can react by nitrous-acid-catalysed routes to give the nitro products. One such pathway involves conventional nitrosation followed by oxidation, whilst the other clearly involves radical cation intermediates, and is initiated by a one-electron transfer reaction involving the nitrosonium ion. This mechanism is given in scheme (3.27) in a more generalised and simplified form, leaving out the possible involvement of radical pairs. Either the first stage (*a*) or the last stage (*b*) can be rate determining, depending upon the $[NO^+]$. The second stage is written here in parentheses, since it probably involves a more complex (and as yet unidentified) route.

$$\left.\begin{array}{c} ArH + NO^+ \xrightleftharpoons{(a)} Ar\overset{+}{H}^\bullet + NO^\bullet \\ (NO^\bullet + NO_2^+ \rightleftharpoons NO^+ + NO_2^\bullet) \\ ArH^{+\bullet} + NO_2^\bullet \xrightarrow{(b)} ArNO_2 + H^+ \end{array}\right\} \qquad (3.27)$$

The key step in the mechanism outlined in scheme (3.27) is the one-electron transfer reaction to nitrosonium ion to give radical cation species. This step may be involved more generally than is hitherto believed. Evidence has been presented[43] which suggests that the nitration of naphthalene in nitric acid with added oxides of nitrogen involves the formation of radical cations. Here, however, the situation is more complex, probably involving a chain mechanism since the kinetic order with respect to naphthalene is 1.5.

Nitrosonium ions also catalyse the exchange of N—H protons between N,N-dimethylanilinium ions and the solvent in *ca* 90% sulphuric acid, under conditions where the *C*-nitrosation reaction is much slower than the exchange reaction.[44] The rate equation is the same as that found for the nitrous-acid-catalysed nitration at low $[HNO_2]$ (ie first-order in $[ArNMe_2H^+]$ and also first-order in $[NO^+]$). The proposed mechanism involves the formation of a loose complex between NO^+ and $ArNMe_2H^+$, followed by a reversible loss of the N—H proton. This reaction is clearly closely related (as the first stage in both cases) to the direct diazotisation of anilinium ions which occurs for the more basic anilines, and also to the nitrous-acid-catalysed nitration reaction.

References

1. K. Schofield, *Aromatic Nitration*, Cambridge University Press, Cambridge, 1980.
2. B.C. Challis, R.J. Higgins & A.J. Lawson, *J. Chem. Soc., Perkin Trans. 2*, 1972, 1831.
3. S. Veibel, *Ber.*, 1930, **63**, 1577.
4. E. Kremos, N. Wakeman & R. Hixon, *Org. Synth.*, 1941, **1**, 511.
5. A. Kraus, *Monatsh. Chem.*, 1881, **12**, 371.
6. J. Suzuki, N. Yagi & S. Syzuki, *Chem. Pharm. Bull.*, 1984, **32**, 2803.
7. Y.L. Chow & Z.Z. Wu, *J. Am. Chem. Soc.*, 1985, **107**, 3338.
8. P.B.D. de la Mare & J.H. Ridd, *Aromatic Substitution*, Butterworths, London, 1959, p. 96.
9. H.H. Hodgson & D.E. Nicholson, *J. Chem. Soc.*, 1941, 470.

10. W.J. Hickinbottom, *J. Chem. Soc.*, 1933, 946.
11. J.M. Tedder, *J. Chem. Soc.*, 1957, 4003.
12. O. Baudisch, *Science*, 1940, **92**, 336; G. Cronheim, *J. Org. Chem.*, 1947, **12**, 1, 7, 20.
13. R.A. Henry, *J. Org. Chem.*, 1958, **23**, 648.
14. M. Colonna, L. Greci & M. Poloni, *J. Chem. Soc., Perkin Trans. 2*, 1984, 165.
15. R.G. Coombes in *Comprehensive Organic Chemistry*, Vol. 2, Ed. I.O. Sutherland, Pergamon, Oxford, 1979, p. 310.
16. J.H. Boyer in *The Chemistry of the Nitro and Nitroso Groups*, Ed. H. Feuer, Interscience, New York, 1969, Part 1, pp. 215–99.
17. K. Hafner & K.L. Montz in *Friedel Crafts and Related Reactions*, Vol. 4, Ed. G.A. Olah, Interscience, New York, 1965, p. 164.
18. B.C. Challis & A.J. Lawson, *J. Chem. Soc. (B)*, 1971, 770.
19. B.C. Challis & R.J. Higgins, *J. Chem. Soc., Perkin Trans. 2*, 1973, 1597.
20. K.M. Ibne-Rasa, *J. Am. Chem. Soc.*, 1962, **84**, 4962.
21. A. Castro, E. Iglesias, J.R. Leis, M. Mosquera & M.E. Pena *Bull. Soc. Chim. Fr.*, 1987, 83.
22. B.C. Challis & R.J. Higgins, *J. Chem. Soc., Perkin Trans. 2*, 1972, 2365.
23. L.R. Dix & R.B. Moodie, *J. Chem. Soc., Perkin Trans. 2*, 1986, 1097.
24. B.C. Challis & A.J. Lawson, *J. Chem. Soc., Perkin Trans. 2*, 1973, 918.
25. B.C. Challis & R.J. Higgins, *J. Chem. Soc., Perkin Trans. 2*, 1975, 1498.
26. K.E. Spratley, Ph.D. Thesis, The City University, 1982.
27. S. Uemura, A. Toshimitsu & M. Okano, *Bull. Chem. Soc. Jpn*, 1976, **49**, 2582.
28. V.I. Minkin, R.M. Minyaev, I.A. Yudilevich & M.F. Kletskii, *Zh. Org. Khim.*, 1985, **21**, 926.
29. W.D. Reents & B.S. Freiser, *J. Am. Chem. Soc.*, 1980, **102**, 271.
30. H.H. Hodgson & A. Kershaw, *J. Chem. Soc.*, 1930, 277.
31. C.L. Perrin, *J. Am. Chem. Soc.*, 1977, **99**, 5516; L. Eberson & F. Radner, *Acta Chem. Scand. Ser. B*, 1984, **B38(10)**, 861.
32. H. Martinsen, *Z. Phys. Chem.*, 1904, **50**, 385; F. Arnall, *J. Chem. Soc.*, 1923, **123**, 3111; 1924, **125**, 811.
33. J.G. Hoggett, R.B. Moodie & K. Schofield, *J. Chem. Soc. (B)*, 1969, 1.
34. C.A. Bunton, E.D. Hughes, C.K. Ingold, D.I.H. Jacobs, M.H. Jones, G.J. Minkoff & R.I. Reed, *J. Chem. Soc.*, 1950, 2628.
35. J.C. Giffney & J.H. Ridd, *J. Chem. Soc., Perkin Trans. 2*, 1979, 618.
36. A.H. Clemens, J.H. Ridd & J.P.B. Sandall, *J. Chem. Soc., Perkin Trans. 2*, 1984, 1667.
37. D.J. Mills, Ph.D. Thesis, University of London, 1976; F. Al-Omran, K. Fujiwara, J.C. Giffney, J.H. Ridd & S.R. Robinson, *J. Chem. Soc., Perkin Trans. 2*, 1981, 518.
38. D.S. Ross, G.P. Hum & W.G. Blucker, *J. Chem. Soc.,Chem. Commun.*, 1980, 532.
39. M. Ali & J.H. Ridd, *J. Chem. Soc., Perkin Trans. 2*, 1986, 327.
40. U. Al-Obaidi & R.B. Moodie, *J. Chem. Soc., Perkin Trans. 2*, 1985, 467.

41. L. Main, R.B. Moodie & K. Schofield, *J. Chem. Soc., Chem. Commun.*, 1982, 48.
42. A.H. Clemens, J.H. Ridd & J.P.B. Sandall, *J. Chem. Soc., Perkin Trans. 2*, 1984, 1659.
43. D.S. Ross & K.D. Moran, *J. Org. Chem.*, 1983, **48**, 2118.
44. D.J. Mills & J.H. Ridd, *J. Chem. Soc., Perkin Trans. 2*, 1980, 637.

4

N-Nitrosation

By far the best-known and most widely studied nitrosation reactions are *N*-nitrosations of amines and related compounds. Many of these reactions have been known for a long time, indeed one of the very first examples reported, involved the conversion of aspartic acid into malic acid in 1846 by Piria;[1] deamination of simple aliphatic and also of aromatic primary amines was achieved soon thereafter.[2] *N*-Nitrosations occur widely in organic chemistry in many standard synthetic pathways, and a number (including diazotisation and azo dye formation) have large scale industrial applications. Because of their importance, such reactions have been also much studied mechanistically, and in general are now well understood. Many reactive intermediates in nitrosation have been identified in kinetic studies, using amine (or closely related) substrates. It is worth noting that a large range of mechanistic features can appear in nitrosation. These include acid catalysis, base catalysis, nucleophilic catalysis, reversibility, encounter-controlled reactions, intramolecular rearrangements, C—C bond cleavage reactions, and *ipso*-attack. Many of these have been demonstrated unambiguously in reactions involving attack at the nitrogen atom. The whole area has been a particularly fruitful one for mechanistic investigations.[3,4] Since the discovery that nitrosamines are powerful carcinogens in all animal species which have been tested, the nitrosation of secondary (and tertiary) amines has been very widely studied, particularly from the viewpoint of the possible *in vivo* formation of nitrosamines from naturally occuring secondary amines and sources of nitrous acid in foods and in water supplies.

4.1 Primary amines

4.1.1 Products

Both aliphatic and aromatic primary amines react readily with a range of nitrosating agents to give finally, products of deamination, including alcohols, alkenes, alkyl and aryl halides and phenols. The first, and generally rate-limiting step is one of N-nitrosation, for both aliphatic and aromatic amines, which results in the formation of a primary nitrosamine RNHNO as outlined in scheme (4.1). In most cases these nitrosamines are not stable (contrasting with the nitrosamines derived from secondary amines), and in a series of rapid reactions, including proton transfer and the loss of a water molecule, give the diazonium ion RN_2^+ as shown in scheme (4.2). Diazonium ions, although they can be isolated as salts in some aromatic cases, are themselves generally reactive species whose reactions will be discussed later. The corresponding reactions of secondary amines (again both aliphatic and aromatic) stop at the nitrosamine stage, since there are now no α-hydrogen atoms available for the necessary proton transfer reactions leading to diazonium ion formation. Nevertheless, since the rate-limiting process is the same for primary and secondary amines, the kinetic equations are very similar, and the rate constants are also often not very different, eg as in the case of the nitrous acid reaction with aniline and N-methylaniline.[5]

$$RNH_2 \xrightarrow{\text{XNO}} R\overset{+}{N}H_2NO \longrightarrow RNHNO \tag{4.1}$$

$$RNHNO \rightleftharpoons RN{=}N{-}OH \xrightarrow{\text{H}^+} RN{=}N{-}\overset{+}{O}H_2 \longrightarrow RN_2^+ + H_2O \tag{4.2}$$

4.1   **4.2**

Primary aromatic nitrosamines have been detected spectroscopically at low temperatures,[6] whilst the aliphatic counterparts appear to be so reactive that they cannot be observed directly, and their presence in diazotisation is therefore inferred. However, some heterocyclic primary amines give rise to particularly stable primary nitrosamines such as the two examples (from an amino triazole and an amino oxadiazole) shown in **4.1** and **4.2**. These can readily be isolated from very dilute acid solution and characterised; their unusual stability has been attributed[7] to internal hydrogen-bonding involving the diazohydroxy form of the nitrosamine (see equation (4.3)). When treated with stronger acid they form the corresponding diazonium ions.

$$\tag{4.3}$$

All of the usual nitrosating agents discussed in chapter 1 will nitrosate amines. The most commonly used preparative procedures are those using aqueous acidic nitrous acid or alkyl nitrites in non-aqueous solvents.

(a) Aromatic amines

Although aryl diazonium ion salts such as the tetrafluoroborate and hexafluorophosphate, can be isolated, the ions are most frequently prepared and used in solution, since the solid salts are mostly explosively unstable. The reactions of diazonium ions are very well known in organic chemistry, and the species are important intermediates in many synthetic pathways. A few examples are shown here in equations (4.4)–(4.8), to illustrate the diversity and importance of these reactions. These examples are by no means comprehensive, but there are a number of excellent books and review articles available,[8] which discuss diazotisation and the reactions of diazonium ions in detail.

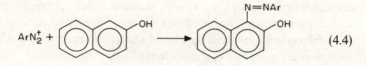

$$ArN_2^+ + RNH_2 \longrightarrow ArN{=}N{-}NHR \tag{4.5}$$

$$ArN_2^+ + Cu^I Cl \longrightarrow ArCl \tag{4.6}$$

$$ArN_2^+ + Y^- \longrightarrow ArY + N_2 \tag{4.7}$$

$$ArN_2^+ BF_4^- \xrightarrow{\text{heat}} ArF + BF_3 + N_2 \tag{4.8}$$

Equation (4.4) is an example of the formation of azo dyes by electrophilic aromatic substitution by the diazonium ion, and is one of the very important reactions involved in dye production on a large industrial scale. Substituents both in the aromatic substrate and in the diazonium ion generate new products, often with important colouring properties. Electrophilic attack (by the diazonium ion) can also occur at the nitrogen atom of eg an amine (equation (4.5)) to give triazene products. Equation (4.6) is an example of the well-known Sandmeyer reaction, using Cu(I) halides (and other salts) to produce halo- and other substituted aromatic compounds. Aryl bromides, iodides and thiocyanates as well as chlorides can be made in this way; the reaction mechanism here is quite complex and is believed to involve radical intermediates. The dediazoniation reaction (replacement of N_2^+ by a nucleophile Y^-, equation (4.7)) occurs readily for a range of nucleophiles eg water, chloride ion, flouride ion, bromide ion, thiocyanate ion. Three distinct reaction pathways have been established, (a) a unimolecular reaction involving rate-limiting formation of an aryl cation and nitrogen, (b) a bimolecular mechanism involving synchronous attack by Y and loss of nitrogen, and (c) an elimination–addition pathway, where the intermediate is a carbene. Interestingly an exchange of nitrogen atoms accompanies the hydrolysis of the diazonium ion (equation (4.9)).[9] This is believed to occur via an intermediate where the aryl cation and molecular nitrogen are held together in a solvent cage, allowing recapture to occur before the species are separated by the solvent. The dediazoniation reactions of aryl diazonium ions have been thoroughly reviewed by Zollinger.[10] Equation (4.8) is the Schiemann reaction, a convenient preparative method for the production of aryl fluorides in high yield.

$$Ph\!-\!{}^{15}\overset{+}{N}\!\!\equiv\!\!N \rightleftharpoons Ph\!-\!\overset{+}{N}\!\!\equiv\!\!{}^{15}N \qquad (4.9)$$

Many other reactions of aryl diazonium ions are known, including nucleophilic attack at the terminal nitrogen atom, coupling with aliphatic carbon systems, and the formation of arene, biaryl and azoarene products. Aryl diazonium ions are also a good source of aryl radicals, which can be used to effect arylation of aromatic (and other compounds). They can also generate heterocyclic compounds by cyclisation and also can be used to bring about both ring expansion and ring contraction reactions. The equilibria involving diazohydroxide and diazotate formation from diazonium ions and hydroxide ion (equation (4.10)) have also been much studied.[11] Saunders reported a systematic classification of the reactions of diazonium ions in 1949[12] and this has been brought up to date by Wulfman in 1978.[13]

$$ArN_2^+ + OH^- \rightleftharpoons ArN\!=\!N\!-\!OH \overset{OH^-}{\rightleftharpoons} ArN\!=\!N\!-\!O^-$$
$$\qquad (4.10)$$

(b) Aliphatic amines

It is to be expected that aryl diazonium ions should be much more stable than their aliphatic counterparts, since the positive charge formally on nitrogen can be delocalised into the aromatic systems. Heterocyclic diazonium ions can be similarly stabilised. The absence of such stabilisation in aliphatic diazonium ions makes them very unstable towards loss of nitrogen, forming the carbocation. In a few special cases, where there are powerful electron-withdrawing groups present (eg as in the hexafluoropropane-2-diazonium ion[14]), the aliphatic diazonium ions have been detected spectroscopically. Presumably the trifluoromethyl groups destabilise the carbocation. In general, however, there is no chemistry of aliphatic diazonium ions (such as the azo coupling reaction) to compare with that of aryl diazonium ions. However, the subsequent reactions of the carbocations derived from primary aliphatic amines have been much studied, and have presented something of a problem in the understanding of the mechanism involved, particularly in comparison with the reactions of carbocations generated eg by solvolysis.

The so-called deamination reaction proceeds readily with most aliphatic primary amines using aqueous acidic nitrous acid, or with alkyl nitrites in acetic acid. Usually a range of products is formed, including alkenes, alcohols and in the case of the acetic acid reactions, acetates. Further, products are also encountered in which skeletal rearrangement has occurred. These reactions are set out in scheme (4.11) in general, and in terms of the expected simple reactions of carbocations, ie nucleophilic attack by the solvent and elimination of a proton, both with and without alkyl and hydrogen atom rearrangement reactions.

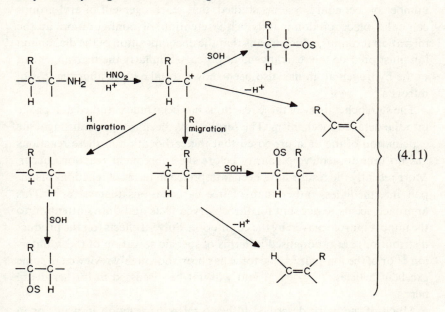

(4.11)

(SOH = hydroxylic solvent, eg H_2O, CH_3COOH)

The long-standing problem with these deamination reactions is that the product distribution has often been very different from that observed when the carbocations are generated by other reactions eg by solvolysis. The migratory aptitude of eg aryl groups is often very much smaller than it is in solvolysis. Initially two explanations were offered; (a) that the carbocations generated in deamination were of very high energy ('hot' carbocations), which would not be very discriminating in their subsequent reactions,[15] and (b) that the branching point for product formation is not the carbocation itself, but its immediate precursor, the diazonium ion,[16] this would mean that the deamination and solvolysis reactions were not comparable.

The concept of 'hot' carbocations is now largely discounted in the context of these reactions. One test of the hypothesis concerned a comparison of the distribution of deuterium labelling in a carbocation generated by both routes.[17] The distribution was found to depend only upon the nucleophilicity of the solvent, and not upon the nature of the reaction in which the carbocation is generated.

The idea that the diazonium ion (and not the carbocation) is the reactive intermediate is an attractive one and explains a number of the discrepancies between the two reactions. However, Collins[18] has shown, at least for a number of secondary systems studied, that rearrangement of aryl groups can take place predominantly with retention of configuration at the migration terminus, which means that the decomposition of the diazonium ion must precede the rearrangement process. Similarly the rearrangement of the hydrogen atom has also been shown to take place after the loss of nitrogen.[19]

The stereochemistry of these reactions has been much studied as a probe into the detailed mechanism. The results have been rather confusing. One explanation of the data proposed that the carbocation in these reactions has no time to achieve planarity before the subsequent reactions occur. More recently the results have been interpreted in terms of reactions of ion-pair intermediates, rather than of free carbocations themselves.[20] This approach seems to account for the observed facts and draws attention to the important role played by the counterion. Solvent effects (on the product distribution) are rationalised in terms of specific solvation of the counter-ion,[21] or of the ion-pair.[22] The topic has been thoroughly reviewed in some excellent articles,[20,23,24,25,26] and will not be discussed in further detail here.

Much of the work described hitherto refers to reaction in water or in acetic acid solvent. More recently the range of both nitrosating agents and also of solvents has been extended, in all cases in attempts to improve yields and also to reduce the range of products formed. There is a full survey[27] of the study of the deamination of aliphatic amines in a range of aprotic solvents. Here again, as in protic solvents, the products include alkyl halides, alkenes, rearrangement products, products of solvent displacement and also diazoalkanes. In some cases the use of aprotic solvents results in a simpler product composition, and so the reaction is of more use synthetically. Some examples are given in equations (4.12), (4.13) and (4.14) to illustrate the range of conditions available. The use of nitrosonium salts in nitromethane solvent is particularly suitable for using the carbocation to alkylate aromatic systems (equation (4.12)) and trimethylsilyl chloride with an alkyl nitrite in methylene chloride and other solvents is a good method for alkyl chloride formation (equation (4.14)). These and other procedures are well covered in review articles.[20,28,29]

$$ArH + RNH_2 + NO^+PF_6^- \xrightarrow{\text{nitromethane}} ArR + N_2 + H_2O + HPF_6 \quad (4.12)$$

$$RNH_2 + RONO + HOAc \xrightarrow{\text{chloroform}} \text{Hydrocarbon products} \quad (4.13)$$

$$RNH_2 + RONO + (CH_3)_3SiCl \xrightarrow[\text{solvents}]{\text{aprotic}} RCl \quad (4.14)$$

4.1.2 Kinetics

Kinetic measurements of the nitrosation of primary amines (particularly aromatic amines) have played a particularly important part in the establishment of nitrosation mechanism generally. Apart from giving structure–reactivity correlations, these studies have been particularly useful in identifying the specific nitrosating agents involved, over a range of experimental conditions. The early literature produced a number of apparently conflicting results, particularly with regard to the overall kinetic order of the reaction. These and later results were resolved into a consistent mechanistic framework, mainly in a series of papers[30] by Hughes, Ingold and Ridd in 1958. The position was summarised in an excellent review article[3] by Ridd in 1961, and although a number of developments have occurred since that time, the mechanistic framework established then remains valid.

(a) Low acid concentration

The reaction mechanism is crucially dependent upon the acidity of the medium and it is useful to discuss three broad acid ranges. In dilute (up to *ca* 0.1 M) acid solution nitrous acid reacts with primary amines by two pathways, one involving dinitrogen trioxide N_2O_3 as the reactive species, and the other the nitrous acidium ion $H_2NO_2^+$ (or the free nitrosonium ion NO^+ which is kinetically indistinguishable from the nitrous acidium ion, see section 1.1). These two mechanisms are characterised by the two rate equations (4.15) and (4.16) for reaction in water. The pathway via dinitrogen trioxide is favoured at low acidity and at high $[HNO_2]$, and also for the more basic aliphatic amines. The mechanisms showing nitrosation via dinitrogen trioxide and nitrosation via nitrous acidium ion (or nitrosonium ion) are given in schemes (4.17) and (4.18). Given the equilibrium constant for dinitrogen trioxide formation[31] (for a discussion of this point see section 1.1) it is easy to obtain the bimolecular rate constant for attack of dinitrogen trioxide. Such values have been obtained for a large number of primary (and secondary) amines, some of which are included in table 1.1. It is clear that dinitrogen trioxide is a very reactive species and for many amines the rate constant is close to that calculated for an encounter-controlled process. This is also indicated by the relative insensitivity of the rate constant, for the more reactive species, to the substrate structure. Further the activation energy in these cases is much smaller than that expected for a 'chemically-controlled' process.

$$\text{Rate} = k[RNH_2][HNO_2]^2 \qquad (4.15)$$

$$\text{Rate} = k[RNH_2][HNO_2][H^+] \qquad (4.16)$$

$$\left. \begin{array}{c} HNO_2 + H^+ \rightleftharpoons H_2NO_2^+ \\ (\text{or } HNO_2 + H^+ \rightleftharpoons NO^+ + H_2O) \\ H_2NO_2^+ \ (\text{or } NO^+) + NO_2^- \rightleftharpoons N_2O_3 + H_2O \\ N_2O_3 + RNH_2 \longrightarrow RN_2^+ \end{array} \right\} \qquad (4.17)$$

$$\left. \begin{array}{c} HNO_2 + H^+ \rightleftharpoons H_2NO_2^+ \\ (\text{or } HNO_2 + H^+ \rightleftharpoons NO^+ + H_2O) \\ H_2NO_2^+ \ (\text{or } N\overset{+}{O}) + RNH_2 \longrightarrow RN_2^+ \end{array} \right\} \qquad (4.18)$$

For the more reactive amines, particularly at low acid concentrations and at high amine concentrations (where the concentration of the free amine is greatest), the reaction becomes zero-order in the amine (ie rate equation (4.19) now applies). This is interpreted in terms of the same sequence of reactions outlined in scheme (4.17) except that the rate of formation of dinitrogen trioxide is now rate-limiting (ie the rate of reaction of dinitrogen trioxide with the amine is now much faster than its hydrolysis to nitrous acid). This limiting condition has been achieved with aniline and hydroxylamine as well as for a number of other amine and non-amine substrates.

$$\text{Rate} = k[HNO_2]^2 \qquad (4.19)$$

Since the equilibrium constant for nitrous acidium ion formation has not been directly determined it is not possible for the mechanism outlined in scheme (4.18) to determine the bimolecular rate constant for the reaction. However the third-order rate constant given in equation (4.16) has been determined for many amine (and other) reactants, and its value gives a measure of the reactivity of the amine. Some values are given in table 4.1. There are no data for the more basic aliphatic amines, which react preferentially with dinitrogen trioxide. There is a tendency for the k values in table 4.1 to approach an upper limit at $25\,°C$ of *ca* $6000\ l^2\,mol^{-2}\,s^{-1}$. This limit also occurs with other substrates, and it has been argued[38] that this is the encounter limit for reaction between the amine (and other substrates) and the actual nitrosating agent. If the latter is the nitrous acidium ion then assuming that the bimolecular rate constant limit is $7 \times 10^9\ l\,mol^{-1}\,s^{-1}$, then the estimated pK_a of the nitrous acidium ion is *ca* -6.

It should be noted that k in equation (4.16) refers to reaction with the free base form of the amine. All the data in table 4.1 were obtained for reaction in acid solution where the degree of amine protonation is often extensive. Thus if the rate equation is expressed in terms of the total stoichiometric concentration of the amine $[RNH_2]_T$, then the rate equation takes the

Table 4.1 *Values of k in Rate $= k[HNO_2][RNH_2][H^+]$ for some primary amines*

Amine	$k/l^2\,mol^{-2}\,s^{-1}$	Temperature/°C	Reference
Aniline	4600	25	32
2-Methylaniline	4500	25	32
4-Methylaniline	6000	25	32
4-Chloroaniline	4200	25	32
Sulphanilic acid	7300	25	33
4-Nitroaniline	2700, 161	25, 0	32, 34
Sulphanilamide	900	25	33
2, 4-Dinitroaniline	$2.5, 0.37^a$	25, 0	33, 34
Sulphamate ion	1167, 170	25, 0	35, 36
Sulphamic acid	0.96	0	36
Phenylhydrazine	870	25	37
4-Methylphenyl- hydrazine	1400	25	37

^{a}The value often quoted in review articles is $3.7\ l^2\,mol^{-2}\,s^{-1}$, but there is an error in the original paper and the correct value is $0.37\ l^2\,mol^{-2}\,s^{-1}$.

modified form given in equation (4.20) where K_a is the dissociation constant of protonated amine. A similar correction must also be applied for the ionisation of nitrous acid to nitrite ion (pK_a 3.2), if reactions are carried out at pH > ~ 2.

$$\text{Rate} = \frac{kK_a[\text{RNH}_2]_T[\text{HNO}_2][\text{H}^+]}{K_a + [\text{H}^+]} \tag{4.20}$$

Both corrections of course also apply to the dinitrogen trioxide reactions, and the full rate equation is given in equation (4.21), in which $K_{\text{N}_2\text{O}_3}$ is the equilibrium constant for dinitrogen trioxide formation and K_N is the dissociation constant for nitrous acid. At a low acidity therefore, increasing the acidity increases the concentration of molecular nitrous acid whilst decreasing the concentration of the free base form of the amine. The nett result is that the measured coefficient goes through a maximum value at a pH value which is governed by the basicity of the amine. Similarly the variation with acidity of the measured rate coefficient for reaction via nitrous acidium ion (or nitrosonium ion), depends on the basicity of the amine. For amines with a low pK_a (eg 2,4-dinitroaniline) the reaction is acid-catalysed over a large acidity range, whereas for the more basic amines, acid catalysis occurs only at very low acidities, and disappears at higher acid concentration when $[\text{H}^+] \gg K_a$ in equation (4.20).

$$\text{Rate} = \frac{kK_{\text{N}_2\text{O}_3}K_a[\text{RNH}_2]_T[\text{HNO}_2]^2[\text{H}^+]^2}{(K_a + [\text{H}^+])(K_\text{N} + [\text{H}^+])^2} \tag{4.21}$$

(b) *Moderate acidities*

At higher acidities (0.1–6.5 M perchloric acid) another mechanism of nitrosation of aromatic amines comes into play, which is not found in aliphatic amines. The rate equation is now of the form of equation (4.22), and has been established for aniline derivatives with $pK_a > ~ 3$. The striking feature is that now marked acid catalysis occurs even though $[\text{H}^+] \gg K_a$. This has been explained by Ridd[39,40] in terms of a mechanism where the reactive species is now the protonated form of the amine and not the free base form. The detailed form is set out in scheme (4.23) and involves the rapid reversible formation of a complex (maybe a π complex) between the protonated amine and nitrosonium ion, followed by a rate-limiting rearrangement of the nitroso group to the amino nitrogen atom, which occurs simultaneously with proton transfer to the solvent S. Complex formation of this type is well known with aromatic systems in strong acid solution, but would not occur with aliphatic amines. The strongest evidence for this mechanism, which rather unusually involves reaction between two positively charged species, comes from the kinetic substituent effects. These effects, which are quite different from those found when the reagent is dinitrogen trioxide or a nitrosyl halide, do not parallel the basicity trend of the anilines, but rather are better understood in terms of their effect on the acidity of the N-H protons as required by the mechanism in scheme (4.23). In particular the significant activating effects of electron-donating substituents in the 3- and 4-positions differentiate clearly between the reactions of the free base form and the protonated form of the anilines.

$$\text{Rate} = k[\text{ArNH}_3^+][\text{HNO}_2]h_0 \tag{4.22}$$

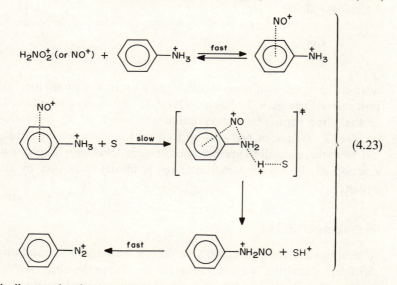

$$(4.23)$$

A similar mechanism occurs in reactions between aniline derivatives and nitrosamines, again a diazotisation reaction where the nitrosamine can be thought of as a carrier of nitrosonium ion. Again the rate equation suggests a direct reaction between the protonated aniline and the nitrosamine[41] as indicated in equation (4.24). There is a striking resemblance between the kinetic substituent effects in this reaction and those in the nitrous acid diazotisation as shown in table 4.2.

$$(C_6H_5)_2\overset{+}{N}HNO + Ar\overset{+}{N}H_3 \longrightarrow (C_6H_5)_2NH + Ar\overset{+}{N}H_2NO + H_3O^+$$

$$(4.24)$$

Nitrosation of both hydrazine and alkyl- and aryl-substituted hydrazines[37,42] is also believed to involve a direct reaction between a positive

Table 4.2. *Substituent effects in the nitrosation of anilinium ion by nitrous acidium ion/nitrosonium ion and the protonated form of N-nitrosodiphenylamine*

| | Relative rate constants | |
	$H_2NO_2^+/NO^+$	$Ph_2\overset{+}{N}HNO$
Aniline	1.0	1
4-Methylaniline	7.4	9.1
3-Methylaniline	6.8	–
4-Chloroaniline	0.2	0.8
3-Methoxyaniline	18.6	4.4

nitrosating species, nitrous acidium ion or nitrosonium ion, and the protonated form of the hydrazine $RNH\overset{+}{N}H_3$. Similarly a reaction pathway in the nitrosation of hydroxylamine and some of its derivatives[43,44] involves a reaction of $\overset{+}{N}H_3OH$ with nitrous acidium ion or nitrosonium ion. Both these reactions are discussed in more detail later in section 4.5, but reaction of the protonated substrates is clearly indicated by the rate equations.

(c) Concentrated acids

At acidities greater than *ca* 60% perchloric acid and 60% sulphuric acid, nitrous acid is virtually quantitatively converted to the nitrosonium ion.[45] In this region the rate of diazotisation of anilines decreases with increasing acidity according to rate equation (4.25).[46] The suggested mechanism given in scheme (4.26) involves the rapid and reversible formation of the protonated primary nitrosamine which undergoes a rate-limiting proton transfer to the solvent(S) before the series of rapid reactions leading to the diazonium ion. The decreasing rate of proton transfer with increasing acidity is readily explicable in terms of the reduction in the water activity in this region, and is known for other systems.[47] A large kinetic solvent isotope effect ($k_H/k_D \sim 10$) is thought to arise partly from the rate-limiting proton transfer and partly from the isotope effect on the equilibrium position in the initial *N*-nitrosation. Further the effect of substituents in the aromatic ring on the position of equilibrium in the *N*-nitrosation and on the rate of proton transfer should tend to oppose each other, resulting in a very small overall substituent effect, as is found experimentally.

$$\text{Rate} = k[\text{Ar}\overset{+}{\text{N}}\text{H}_3][\text{HNO}_2]h_0^{-2} \tag{4.25}$$

$$\left.\begin{array}{l} \text{ArNH}_3^+ + \text{NO}^+ \underset{}{\overset{\text{fast}}{\rightleftharpoons}} \text{Ar}\overset{+}{\text{N}}\text{H}_2\text{NO} + \text{H}^+ \\[3mm] \text{Ar}\overset{+}{\text{N}}\text{H}_2\text{NO} + \text{S} \xrightarrow{\text{slow}} \text{ArNHNO} + \text{SH}^+ \\[3mm] \text{ArNHNO} \xrightarrow{\text{fast}} \text{ArN}_2^+ \end{array}\right\} \tag{4.26}$$

The reduction in the rate of diazotisation at very high acidities may well provide the explanation as to why some of the more basic aromatic amines undergo *C*-nitrosation instead of diazotisation under these conditions.[48] There is now the possibility of a direct competition between the slow proton transfer and the intramolecular Fischer–Hepp rearrangement of the nitroso group to the 4-position in the ring (see equation (4.27)).

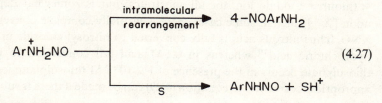

$$\tag{4.27}$$

(d) Nucleophilic catalysis

A feature of the nitrosation of amines is that catalysis occurs in the presence of certain non-basic nucleophiles. The early work concerned halide ion catalysis (chloride ion < bromide ion < iodide ion) but more recently catalysis has also been shown to occur with thiocyanate ion, thiosulphate ion, thiourea, alkyl thioureas and dimethyl sulphide. Many of these features were established using primary amines (particularly anilines) but secondary amines and other substrates show the same effects.

In all cases catalysis arises by the formation of a nitrosyl species XNO (from a nucleophile X^- and acid solutions of nitrous acid), which acts as the nitrosating species usually in a rate-limiting process as outlined in scheme (4.28). For the neutral sulphur nucleophiles X the corresponding nitrosyl species are $X\overset{+}{\text{N}}\text{O}$. These nitrosyl compounds can be detected in solution and the equilibrium constants for their formation, K_{XNO}, have been determined independently in most cases. If reaction occurs with the free base form of the aniline, then the expected rate equation from scheme (4.28) is equation (4.29). This presumes that the extent of conversion of nitrous acid to XNO is small, which is generally but not always the case. In equation (4.29) $[\text{RNH}_2]_\text{T}$ is the total stoichiometric concentration of the amine, and $[\text{HNO}_2]$ the total added nitrous acid concentration. Since for most amines under the experimental conditions used $[H^+] \gg K_a$, there is no overall acid catalysis for these nucleophilic-catalysed reactions. Usually this mechanism accompanies that involving nitrous acidium ion or nitrosonium ion, so that plots of the measured rate constant vs. X^- will be linear with often a small positive intercept.

$$\left. \begin{array}{c} HNO_2 + H^+ + X^- \underset{}{\overset{\text{fast}}{\rightleftharpoons}} XNO + H_2O \\[2ex] XNO + RNH_2 \xrightarrow{\text{slow}} R\overset{+}{N}H_2NO \xrightarrow{\text{fast}} RN_2^+ \end{array} \right\} \qquad (4.28)$$

$$\text{Rate} = \frac{k_2 K_{XNO}[HNO_2][H^+][X^-][RNH_2]_T K_a}{([H^+] + K_a)} \qquad (4.29)$$

The catalytic efficiency usually follows the sequence chloride ion $<$ bromide ion $\sim$ dimethyl sulphide $<$ thiosulphate ion $<$ thiocyanate ion $<$ thiourea $<$ iodide ion, for all substrates, but is somewhat dependent upon $[X^-]$ since in some cases there is virtually complete conversion to XNO. Thus nitrous acid is fully converted to nitrosyl chloride in *ca* 5 M hydrochloric acid[49] whereas in 0.1 M acid full conversion to nitroso-thiosulphate occurs in the presence of 1×10^{-3} M thiosulphate ion.[50] An appropriate correction to equation (4.29) can be made if there is substantial conversion to XNO.

Since K_{XNO} has been measured independently (spectrophotometrically) in most cases, it is easy to obtain values of k_2 (equation (4.29)), the bimolecular rate constant for attack by XNO, usually from plots of the measured rate constant against $[X^-]$. Analysis of the data shows clearly that the catalytic order of reactivity of X^- is governed principally by the magnitude of K_{XNO} (see section 1.2). In fact the actual intrinsic reactivity of the XNO species changes in the opposite sense ie nitrosyl chloride $>$ nitrosyl bromide $>$ nitrosyl thiocyanate $>$ S-nitrosothiouronium ion. Data are presented in tables 1.6, 1.7 and 1.8. The reactivity sequence has been demonstrated for a range of substrates including primary amines, and a more detailed discussion of nucleophilic catalysis is given in section 1.2.

It is also clear from the data in tables 1.6 and 1.7 that for nitrosyl chloride and nitrosyl bromide, the rate constant approaches a limit as the reactivity (the pK_a) of a series of aniline derivatives is increased. This trend is shown much more clearly in figure 4.1.[51] Although there is a little scatter, it appears that for both the nitrosyl chloride and nitrosyl bromide results, the k_2 values are virtually constant for anilines with $pK_a > \sim 4$. The limit is also quite close to that calculated ($7.4 \times 10^9 \, l\,mol^{-1}\,s^{-1}$ at 25 °C) for an encounter-controlled reaction between two uncharged species. There are only a few results for nucleophilic catalysis in the diazotisation of primary aliphatic amines; there are many more for secondary aliphatic amines which are discussed in section 4.2, but it does appears that for aliphatic systems the limiting rate constant is smaller by about 10^2 than is the corresponding limit for aromatic systems . This is discussed more fully in section 4.2. For nitrosyl thiocyanate and S-nitrosothiouronium ion the data analyse reasonably well as Brønsted plots (figure 4.1). There has been no report where either of these reagents is believed to react at the encounter rate. Here the activation energies are significantly larger than are those for nitrosyl chloride and nitrosyl bromide reactions for the more reactive anilines.

It is worth noting here that for the very reactive aromatic amines at relatively high concentration of the free base form (ie at low acidity), it is possible to change the rate-limiting step to that of the formation of XNO. With aniline itself, this has been achieved for both bromide and iodide ion.[52] This point is discussed more fully in section 1.2.

In some cases, for the less basic aromatic amines, and also at relatively high $[X^-]$, the linear dependence of the measured rate constant upon $[X^-]$ disappears. For example,[53] curved plots have been obtained for the reactions of 4-nitro-, 2-nitro-, 4-carboxy-, 4-chloro- and 3-methoxyanilines, for the bromide-ion-catalysed reactions. This is interpreted in terms of the reversibility of the initial N-nitrosation (see scheme (4.30)). This will become important if $k_{-2}[Br^-]$ is comparable with k_3, and is more likely at high $[Br^-]$ and when there are electron-withdrawing groups present (larger k_{-2}). There is a change of rate-limiting step, from the attack by nitrosyl

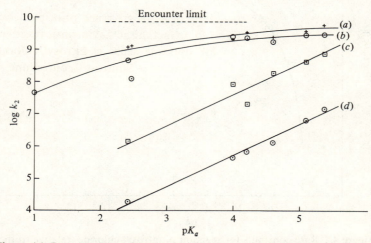

Figure 4.1 Rate constants for reaction of (a) nitrosyl chloride (b) nitrosyl bromide (c) nitrosyl thiocyanate (d) S-nitrosothiouronium ion with aniline derivatives, as a function of pK_a.

bromide to one of the stages in the decomposition $Ar\overset{+}{N}H_2NO \rightarrow ArN_2^+$, which might be a proton transfer, or a loss of a water molecule. The rate equation for the modified scheme is given in equation (4.31) (for $[H^+] \gg K_a$). The kinetic results for a number of amines are quantitatively in agreement with equation (4.31),[53] as shown by the linear double reciprocal plot of (observed rate constant)$^{-1}$ vs. $[X^-]^{-1}$. The same relationship holds for the diazotisation of aniline derivatives and of 1-naphthylamine[54] in 1–5 M hydrochloric acid, when allowance is made for activity effects in these high acid concentrations The $k_{-2}:k_3$ ratios obtained from the double reciprocal plots are *ca* 50 times larger for the bromide-ion-catalysed reactions, than for the chloride ion reactions, reflecting the greater nucleophilicity of bromide ion in aqueous solution.

$$ArNH_2 + XNO \underset{k_{-2}}{\overset{k_2}{\rightleftharpoons}} Ar\overset{+}{N}H_2NO + X^- \qquad (4.30)$$

$$k_3 \downarrow \text{several stages}$$

$$ArN_2^+$$

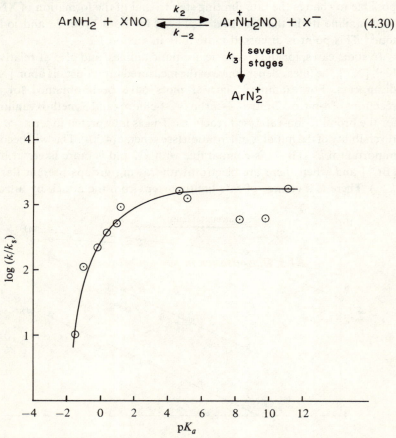

Figure 4.2 Rate constants for reaction of nitrosyl chloride (relative to the hydrolysis reaction) with amines, as a function of pK_a.

$$\text{Rate} = \frac{k_2 k_3 K_{\text{XNO}} K_a [\text{ArNH}_2]_T [\text{X}^-][\text{HNO}_2]}{k_{-2}[\text{X}^-] + k_3} \tag{4.31}$$

The change of rate-limiting steps is complete in 2,4-dinitroaniline, where no bromide ion or thiocyanate ion catalysis occurs. This is also a feature of the nitrosation of amides, which is explained also in terms of scheme (4.30), and which is discussed in more detail in section 4.4. In this regard 2,4-dinitroaniline (and no doubt other amines of very low basicity) behaves as an amide. The reversibility of the initial nitrosation is a quite general feature and crops up also in both *O*- and *S*-nitrosation.

Nitrosyl chloride, dissolved in a non-acidic aqueous solution, reacts with amines to give the expected products, together with nitrite ion formed by hydrolysis of the nitrosyl chloride.[55] These are competitive reactions (see scheme (4.32)) and the ratio of the products $[\text{RN}_2^+]/[\text{NO}_2^-]$ (for primary arylamines) and $[\text{R}_2\text{NNO}]/[\text{NO}_2^-]$ (for secondary amines) has been measured. This ratio gives the ratio of the corresponding rate constants k/k_s ie the rate constant for nitrosation relative to that for hydrolysis. Values of $\log(k/k_s)$ plotted against the pK_a of the amine are shown in figure 4.2, and suggest that, as for reactions where the nitrosyl chloride is generated *in situ*, the reaction is encounter-controlled for amines with p$K_a > \sim 3$, since $\log(k/k_s)$ is approximately constant for p$K_a > \sim 3$.

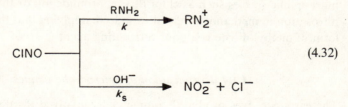

$$\tag{4.32}$$

4.1.3 *Diazotisation in non-aqueous solvents*

On the synthetic side diazotisation can readily be achieved in a range of non-aqueous solvents, including methanol, acetic acid, benzene, ether and acetonitrile, and using a range of nitrosating agents such as alkyl nitrites, nitrosyl halides and nitrosonium salts. These conditions are clearly more appropriate than aqueous conditions if either the amine or the nitrosating agent has a limited solubility in water. Two-phase systems (eg methylene chloride–water) using phase transfer catalysts can also be effective,[56] again where there are solubility problems. This procedure has the advantage of limiting side reactions, and also enables further reactions of diazonium ions (eg azo coupling) to be carried out *in situ*.

The only systems to have been studied mechanistically in a systematic way is that in methanol using sodium nitrite and an acid catalyst.[32] In the presence of hydrogen chloride, reaction occurs via nitrosyl chloride present in very low concentration (as in water). The various equilibria are set out in scheme (4.33). The same reactions occur if the starting reagent is the alkyl nitrite[57] (see section 6.5.2)). Schmid and coworkers established the rate equation (equation (4.34)) for the diazotisation of aniline. This treatment allows for the non-ideal behaviour in relatively high hydrogen chloride concentrations, and $f_\pm$ is the mean ionic activity coefficient of hydrogen chloride in this solvent.

$$\left.\begin{aligned} \mathrm{NO_2^-} + \mathrm{HCl} &\rightleftharpoons \mathrm{HNO_2} + \mathrm{Cl^-} \\ \mathrm{HNO_2} + \mathrm{CH_3OH} &\rightleftharpoons \mathrm{CH_3ONO} + \mathrm{H_2O} \\ \mathrm{CH_3ONO} + \mathrm{HCl} &\rightleftharpoons \mathrm{CH_3OH} + \mathrm{ClNO} \end{aligned}\right\} \qquad (4.33)$$

$$\mathrm{Rate} = k f_\pm^2 [\mathrm{ArNH_3^+}][\mathrm{CH_3ONO}][\mathrm{Cl^-}] \qquad (4.34)$$

In methanol (and no doubt other alcohol solvents) the tendency for the reversal of the *N*-nitrosation to occur is more marked than it is in water. Indeed the effect was first noted (by curved plots of the observed rate constant vs. $[\mathrm{Br^-}]$) for reactions in methanol[58] for both hydrogen chloride and hydrogen bromide as catalysts. A rate equation analogous to equation (4.31) is obeyed and treated by double-reciprocal plots. In the absence of nucleophilic species such as chloride ion, bromide ion or thiocyanate ion nitrosation in methanol is very slow indeed, implying that the protonated form of methyl nitrite is a poor nitrosating agent.

4.1.4 Diazotisation of heterocyclic amines

The early azo dyes produced commercially involved the diazotisation of aniline derivatives. More recently the range of amines has been extended to include heterocyclic amines, since many of them yield azo dyes with improved colour properties. Some examples which have commercial application are shown in **4.3–4.6** and are thiophene, thiazole, oxazole,

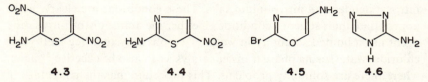

4.3   **4.4**   **4.5**   **4.6**

and triazole derivatives. There is a very large patent literature involving compounds of this type. The amino group here is much less basic than it is in aniline derivatives generally, and protonation often occurs extensively at the ring-nitrogen atoms. On the industrial scale diazotisation is achieved using nitrosyl sulphuric acid in reasonably concentrated sulphuric acid. Nucleophilic catalysis by bromide ion and thiocyanate ion occurs so there may be scope for using somewhat milder conditions.

In some cases the initial nitrosation step is significantly reversible and in others the diazonium ion is very readily hydrolysed, which makes it difficult to use in subsequent reactions. Often the diazonium ions can be stabilised in basic solution by proton loss to give the diazo compound. A good example is the reaction of a substituted amino pyrrole shown in scheme (4.35). Diazotisation of heterocyclic amines is often an important step in the synthesis of some compounds of pharmaceutical value.

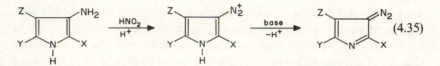

$$\text{(4.35)}$$

Kalatzis and coworkers[59] have studied the kinetics of the reactions of a number of amino pyridine derivatives. For both 2- and 4-aminopyridine protonation occurs at the ring nitrogen (with a basicity comparable with aliphatic amines), and it is the protonated species which reacts with the nitrosating agent (see scheme (4.36)). The situation resembles that of the diazotisation of an aniline derivative containing a strongly electron-withdrawing group. It would be interesting to establish here whether nucleophilic catalysis occurs. It has been suggested that reaction of the nitrosating agent (nitrous acidium ion or nitrosonium ion) occurs first with the heteroaromatic nucleus, followed by a rearrangement of the nitro-sonium ion group to the exocyclic amino group; this rearrangement is believed to occur concurrently with proton loss to the solvent from the amino group. Thus the mechanism is very similar to that proposed for the diazotisation of basic aniline derivatives in moderate acidities discussed earlier (see scheme (4.23)).[40]

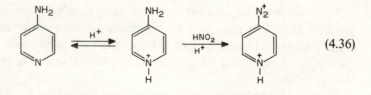

$$\text{(4.36)}$$

4.2 Secondary amines

Secondary amines (both aliphatic and aromatic) react readily with the usual range of nitrosating agents to give nitrosamines (equation (4.37)), which is the same reaction as the first step in the nitrosation of primary amines. However, secondary nitrosamines are relatively stable and cannot be transformed into diazonium salts, since the necessary proton transfer reaction is now not possible. Most reactions are quite rapid and some are significantly reversible (eg with N-acetyltryptophan). The reverse reaction of denitrosation of nitrosamines can be studied if steps are taken to remove XNO (with nitrous acid traps) as it is formed (see section 5.3). Much of what applies to primary amines applies also to secondary amines, particularly the kinetic behaviour, since the rate-limiting step is usually the same in both cases. Nitrosamine formation has, however, assumed an important role ever since the discovery in 1956 that both nitrosamines and nitrosamides are very powerful carcinogens to all animal species which have been tested. This is of some relevance to humans, since nitrosamines can readily be formed *in vivo* from the many naturally occurring secondary amines and sources of nitrous acid – particularly nitrite ion used in food preservation and nitrate ion in water supplies. This aspect is discussed fully in section 5.4. It is not surprising that the nitrosation of secondary amines has been a much studied reaction since 1956.

$$R_2NH + XNO \longrightarrow R_2NNO + HX \qquad (4.37)$$

As for primary amines, the more basic aliphatic secondary amines react with dinitrogen trioxide in nitrous acid solutions. Some of the kinetic results are included in table 1.1. The noteworthy feature is that for a range of 5.6–11.3 in the pK_a values of the amines, the second-order rate constant for dinitrogen trioxide attack is remarkably constant. This suggests that for these amines the reaction is an encounter-controlled process, a conclusion which is borne out by the low energy of activation for this step.[60]

The effect of changing the acidity is the same as for primary amines (see equation (4.21)). For a number of secondary amines the reaction rate constant is at a maximum for the reaction via dinitrogen trioxide at pH 3.0–3.5[61(a),(b)]

Aromatic secondary amines such as N-methylaniline react with dinitrogen trioxide at low acidities but with nitrous acidium ion (or nitrosonium ion) at higher acidities, in both cases in reactions which are encounter-controlled. As expected the reactivity of N-methylaniline is very close to that of aniline.[5]

Nucleophilic catalysis again occurs quite generally and reactions via nitrosyl chloride, nitrosyl bromide, nitrosyl thiocyanate, S-nitroso-thiouronium ion, nitrosothiosulphate ion and nitrosyl acetate have been established kinetically, where these reagents are generated *in situ*. One interesting feature has been noted in a detailed account[62] of the reactions of nitrosyl bromide, and that is, that although aliphatic secondary amines are the more basic, their intrinsic reactivity is significantly ($ca\ 10^2$) less than that of aromatic amines. Nevertheless over a pK_a range of 8.00–11.25 the second-order rate constants for nitrosyl bromide attack lie within the relatively small range 0.7×10^7–$6.2 \times 10^7\ l\,mol^{-1}\,s^{-1}$ at 25 °C. This presents a problem, in that the relative insensitivity of the rate constant to structure implies an encounter-controlled reaction, whilst the actual values are significantly less than the observed limit for the aromatic amine reactions. It may well be that there are two different upper limits to the rate constants for the aliphatic and aromatic systems in some way related to the presence of the π electrons in the latter, although the detailed origin of the effect is not clear at this stage.

Nitrosation of secondary amines can also be readily achieved using nitrite ion in neutral or basic solution, with various carbonyl compounds (eg formaldehyde) acting as catalysts, and also with Fremy's salt. Both of these are discussed in more detail in section 1.8.

Long-chain secondary amines (eg dihexylamine) exhibit micellar catalysis in nitrosation.[63] The extent of catalysis is a function of the chain length and is explained in terms of the electrostatic interactions on the micelle surface and also by preferential solubility of dinitrogen trioxide in the micellar phase.

Secondary amino acids (including sugar amino acids[64]) readily give the corresponding nitrosamines with a range of nitrosating species. Mirvish and coworkers[65] identified kinetically a reaction pathway involving rate-limiting attack by dinitrogen trioxide, but under more acidic conditions nitrous acidium ion (or nitrosonium ion) is the reactive species; nucleophi-

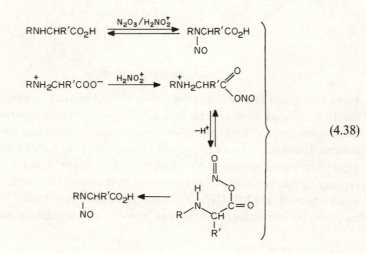

$$(4.38)$$

lic catalysis can also occur. A detailed kinetic study of the reactions of sarcosine and proline[66] revealed an unusual pathway where *O*-nitrosation of the carboxylate ion group occurs first, followed by rate-limiting intramolecular rearrangement of the nitroso group to the nitrogen atom, as outlined in scheme (4.38). This mechanism occurs concurrently with one where the direct attack on nitrogen occurs. The unusual pathway is detected from the dependence of the rate constant for reaction upon the acidity of the medium. As expected, such a pathway is totally absent when the reactant is the ethyl ester of sarcosine.

A similar internal rate-limiting nitroso group rearrangement has been proposed as one of the pathways in the nitrosation of *N*-acetyltryptophan (see scheme (4.39)).[67] Here the initial attack occurs at C-3 (a site known to be prone to electrophilic attack) and internal rearrangement of the nitroso group to the indole nitrogen occurs. Again this mechanism is consistent with the detailed form of the variation of the rate constant with the acidity of the solvent, and may be quite general for indole systems. This mechanism accompanies that where direct nitrosation of the indole nitrogen atom takes place.

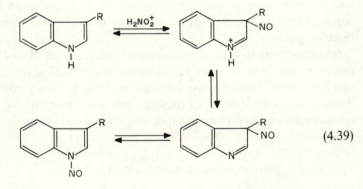

(4.39)

4.3 Tertiary amines

For many years it was generally believed that tertiary amines were not reactive towards nitrous acid. Indeed this view was incorporated into a qualitative test to distinguish between primary, secondary and tertiary amines. However, it had been reported as early as 1864 that triethylamine gave diethylnitrosamine with nitrous acid.[68] Many other examples of tertiary amine nitrosations are reported in the early literature which was summarised in 1963.[69] These reactions have not been studied as much as have the corresponding reactions of primary and secondary amines, and the mechanistic picture is not as clear. More recently the reactions have come under closer scrutiny, particularly since there are a number of tertiary amine structures in drugs, which could develop carcinogenic nitrosamines *in vivo*. Most of the tertiary amine nitrosations studied kinetically, show that the reactions are very much slower than are the corresponding secondary amine reactions (a figure of 10 000 times less reactive has been suggested[61(a)]).

Much of what is known today about tertiary amine nitrosation stems from the work of Loeppky and coworkers. The stoichiometry of the reaction has been established[70] as that given in equation (4.40). A carbonyl compound and nitrous oxide accompany nitrosamine formation. The reaction is sometimes referred to as the nitrosative dealkylation of tertiary amines. Clearly with different substituents in the amine a range of products is possible. A detailed product analysis using the tertiary amines $RC_6H_4CH_2N(CH_2C_6H_5)_2$ was undertaken, and the product radio $[C_6H_5CHO]/[RC_6H_4CHO]$ measured as a function of R. Electronic effects do not seem to affect greatly the position of bond cleavage whereas the size of the alkyl groups α to the nitrogen atom is important. With a bulky alkyl group, the extent of cleavage of that group is much reduced – there is an isotope effect k_H/k_D (measured from product ratios) of 3.8 for the reactions of $R_2NCHR'_2$ and $R_2NCDR'_2$ indicating that the α C—H bond is broken in the rate-limiting step. The formation of nitrous oxide is often associated with the elimination and consequent dimerisation (and a loss of a water molecule) of the nitroxyl radical HNO. All of these results, together with many others in the early literature led Smith and Loeppky[70] to propose the mechanism set out in scheme (4.41). Here the first step is a rapid *N*-nitrosation, which is followed by a rate-limiting elimination of nitroxyl radical to give the imminium ion, which on hydrolysis gives the carbonyl compound and the secondary amine. The latter undergoes a normal rapid nitrosation with more nitrous acid to give the nitrosamine. Later kinetic

$$2R_2NCHR'_2 + 4HNO_2 = 2R_2NNO + 2R'_2CO + N_2O + 3H_2O \quad (4.40)$$

$$
\left.
\begin{array}{l}
R_2N{-}CHR'_2 \;\underset{\text{fast}}{\overset{HNO_2/H^+}{\rightleftharpoons}}\; R_2\overset{+}{N}\underset{NO}{\overset{CHR'_2}{<}} \;\xrightarrow{\text{slow}}\; R_2\overset{+}{N}{=}CR'_2 + HNO \\[2ex]
R_2\overset{+}{N}{=}CR'_2 \;\xrightarrow{H_2O}\; R'_2CO + R_2NH \\[1ex]
R_2NH + HNO_2 \;\xrightarrow{H^+}\; R_2NNO + H_2O \\[1ex]
2HNO \;\longrightarrow\; N_2O + H_2O
\end{array}
\right\} \quad (4.41)
$$

studies[71] showed that there is no catalysis by chloride ion or thiocyanate ion, which suggest that the first stage is in fact a rapid *reversible N-nitrosation*. One major difficulty with kinetic studies here is the fact that the reactions are so slow that the decomposition of nitrous acid becomes a serious side reaction. This may have been overlooked in some studies.

An alternative suggestion has been made[72] as to the fate of the imminium ion. It is suggested that in fact it reacts with nitrite ion (rather than being hydrolysed) to give an unstable alkyl nitrite derivative which rearranges to give a nitrosamine and elimination of a carbonyl compound (see scheme (4.42)). Formation of imminium ions is an integral part of the mechanism of the nitrosation of secondary amines catalysed by carbonyl compounds (see section 1.8), and it is an attractive suggestion that both reactions have this common pathway.

$$\overset{+}{R_2N}{=}CR'_2 + NO_2^- \longrightarrow R_2N{-}CR'_2 \longrightarrow R_2NNO + R'_2CO \qquad (4.42)$$
$$\underset{O{=}N}{\overset{\text{O}}{\longleftrightarrow}}$$

Aromatic dialkylamines when treated with nitrous acid at 80–90 °C in acetic acid give aromatic nitration products as well as those derived from nitrosative dealkylation.[73] Nitro products are believed to arise from nitrogen oxides formed from the decomposition of nitrous acid. The product composition (nitration vs. dealkylation) has been explained in terms of stereoelectronic factors, which control the delocalisation of the unshared electron pair of the amine into the aromatic nucleus.

Interestingly the reactions of some tertiary amines containing other than simple alkyl groups are much faster – although there are no firm kinetic data. This group includes a number of important drugs such as aminopyrine, oxytetracycline and monocycline, and there is some concern regarding the probable formation of nitrosamines under gastric conditions. Different mechanisms may operate here eg the addition of dinitrogen trioxide to the enamine function in aminopyrine; likely possibilities have been discussed.[74]

Nicotine gives rise to nitrosamine products with nitrous acid. *N*-Nitrosonornicotine is the expected product (see equation (4.43)), but other products including nitrosamines are formed from the fragmentation of the pyrrolidine ring.[75]

It has been reported[76] that 2-(*N*, *N*-dimethylaminomethyl) pyrrole reacts with sodium nitrite in acetic acid at 25 °C virtually instantaneously to give dimethylnitrosamine, formaldehyde and a rather unusual oxime derivative (see equation (4.44)). The surprising ease with which this reaction occurs suggests that a mechanism, different from the usual nitrosative dealkylation mechanism (see scheme (4.41)), applies here, but at this stage details are a matter of speculation.

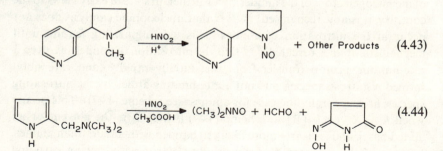

4.4 Amides

Nitrosation of primary amides results in a so-called deamination reaction giving the carboxylic acid and nitrogen as products (equation (4.45)). These reactions are generally much slower than are those of the corresponding amines as expected in view of the presence of the strongly electron-withdrawing carbonyl group. The most suitable reagents for synthetic work have been found to be alkyl nitrites in organic solvents[77] and nitrosonium salts in acetonitrile.[78]

$$RCONH_2 \xrightarrow[\text{or NO}^+]{\text{RONO}} RCO_2H + N_2 \qquad (4.45)$$

Secondary amides react similarly to give the corresponding nitrosamides in a reversible process (see equation (4.46)). To obtain good yields of the nitrosamides it is best to a add a base (eg acetate ion) to remove the acid formed. The base also exerts a catalytic effect. In addition to the reagents mentioned earlier it has been found[79] that nitrosyl chloride and dinitrogen tetroxide in carbon tetrachloride or acetic acid (containing sodium acetate) are efficient reagents. The reaction is quite general and also occurs with ureas and carbamates, and with cyclic derivatives. As for secondary amines there has been a renewed interest in the nitrosation of secondary amides because of the possibility of the formation of carcinogenic nitrosamides within the human body.

$$RCONHR' \underset{\text{or NO}^+}{\overset{\text{RONO}}{\rightleftharpoons}} RCON(NO)R' \qquad (4.46)$$

Initially it was believed that the mechanism of the nitrosation of amides was identical to that of amines. This view derived mostly from kinetic studies by Bruylants and coworkers,[80] and was reviewed in terms of rate-limiting attack by XNO.[81] However, it was later noted that the large range of reactivity encountered is not easily correlated with the known basicities of the nitrogen atom of the amides.[61(a)] Further, in spite of early views to the contrary, it is now recognised[82(a)(b),83] that nucleophilic catalysis does not occur in the nitrosation of amides. This was a puzzling feature until mechanistic studies revealed[83,84] that the nitrosation of amides involved a rate-limiting proton transfer from the initially rapidly (and reversibly) formed *N*-nitroso species, and not a rate-limiting attack by the nitrosating species, as is generally the case for amines (see scheme (4.47)). This occurs when $k_{-1}[\text{X}^-] \gg k_2[\text{B}]$ in scheme (4.47) (where B is the solvent or an added base), and is clearly more likely to happen with amides than amines, because of the presence of a powerfully electron-withdrawing carbonyl group. Under these circumstances, no information can be obtained about the nature of the attacking nitrosating species XNO. Acid catalysis is quite general since protonation of the amide does not occur to any appreciable extent until the acid is very concentrated.

$$
\text{RCONHR'} + \text{XNO} \underset{k_{-1}}{\overset{k_1}{\rightleftharpoons}} \overset{\overset{+}{\underset{|}{\text{NO}}}}{\text{RCONHR'}} + \text{X}^-
$$

$$
\downarrow k_2 \, | \, \text{B}
$$

$$
\underset{\underset{\text{NO}}{|}}{\text{RCONR'}} + \text{BH}^+
$$

(4.47)

The existence of the rate-limiting proton transfer is borne out by the observation of both a primary kinetic isotope effect and general base catalysis.[85] The mechanism is also supported by measurements on the reverse reaction ie the denitrosation of nitrosamides,[82(b),83,84,86] where again there is no nucleophilic catalysis, and the reactions show a primary kinetic isotope effect and general acid catalysis (see section 5.3.1).

It is implied in scheme (4.47) that nitrosation occurs directly at the nitrogen atom of the amide. More detailed studies have suggested that this may not be so.[87] A kinetic analysis of the nitrosation of ethyl-*N*-ethyl-carbamate, *N*, *N'*-dimethylurea and 2-imidazolidine together with some measurements on the reverse reaction (ie denitrosation) has been carried out. An argument based on the nature of the Brønsted plots (slopes and curvatures in some cases) and the use of Eigen theory, shows that the data are not consistent with a direct *N*-nitrosation but rather with *O*-nitrosation, followed by a rearrangement reaction as outlined in scheme (4.48). This is not an unreasonable view; this sort of rearrangement crops up many times in nitrosation, and the fact that protonation of amides[88] and attack by alkylating agents[89] both occur at oxygen lends weight to the argument.

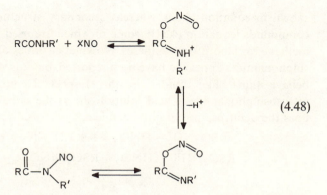

$$\text{(4.48)}$$

Urea itself forms carbon dioxide and nitrogen upon nitrosation (equation (4.49)), and has been much used over the years as a trap or scavenger for nitrous acid. This is necessary, for example, to suppress the nitrosation which occurs in some aromatic nitrations through traces of nitrous acid, in denitrosation reactions (both synthetically and in mechanistic studies), to prevent renitrosation, and to prevent explosive side reactions in the synthesis of alkyl nitrates from alcohols and nitric acid. Indeed these components can be refluxed without difficulty if urea (or some other nitrous acid trap) is present. Urea has the advantage of a high solubility in water and also of forming gaseous products on nitrosation, and so has been much used. Its overall reactivity is, however, quite low and there are many other more efficient scavengers eg hydrazine and azide ion.[33]

$$CO(NH_2)_2 + 2HNO_2 = CO_2 + 2N_2 + 3H_2O \qquad \text{(4.49)}$$

Peptides, in common with amides generally can also be nitrosated. Secondary amino groups give nitrosamides and primary amino groups give products resulting from diazotisation. Reaction of proteins with nitrous acid is the basis of one classical analytical procedure[90] used to determine the number of free amine groups present. *N*-Nitrosopeptides are not very stable and are only isolated and characterised with difficulty.[91] Another study has shown that nitrogen oxides (simulating an urban environment) react with a number of *N*-protected dipeptides to a give dinitroso peptides, mononitroso peptides or a mixture of both, in a way which is governed principally by steric factors rather than by electronic effects.[92] The implication regarding possible dangers from carcinogenic nitrosopeptides derived from ingested nitrite or nitrogen oxides is obvious.

Sulphonamides may also be nitrosated, the primary amino compounds giving the sulphonic acid hydrolysis product and nitrogen (deamination) as shown in equation (4.50), whereas secondary structures give *N*-nitroso compounds (equation (4.51)), some of which are used to generate diazomethane for laboratory syntheses. No kinetic studies of the nitrosation of sulphonamides appear to have been reported, but it is very likely that they behave much like amides in this respect. Denitrosation of a *N*-nitrososulphonamide in acid solution shows the same kinetic features as does the denitrosation of nitrosoamides.[93]

$$RSO_2NH_2 + HNO_2 = RSO_3H + N_2 + H_2O \qquad (4.50)$$

$$RSO_2NHR' + HNO_2 = RSO_2N(NO)R' + H_2O \qquad (4.51)$$

4.5 Other nitrogen-containing compounds

4.5.1 *Ammonia*

The nitrosation of perhaps the simplest nitrogen substrate, although known, has surprisingly not been much studied.[94] It is the presumed reaction when an aqueous solution of ammonium nitrite is heated, in the laboratory preparation of nitrogen (equation (4.52)), and can readily be rationalised as such (see scheme (4.53)). Olah and coworkers[95] have obtained $^{14}N^{15}N$ from $^{15}NO^+BF_4^-$ and ammonia, which is consistent with scheme (4.53) involving the intermediacy of the parent diazonium ion (or protonated nitrogen) HN_2^+. Overall the reaction is very slow in acid solution at 25 °C, because of protonation of ammonia, the reactive species, and so it is difficult to get precise kinetic measurements because of the decomposition of nitrous acid over the necessary time scale, but at low acidity, dinitrogen trioxide has been identified as the effective reagent. Nucleophilic catalysis also occurs,[96] and some results have been obtained[97] for the bromide-ion and thiocyanate-ion-catalysed reactions (with appropriate corrections for the decomposition reaction of nitrous acid). These data show that nitrosyl bromide reacts with ammonia at approximately the same rate as it does with a number of aliphatic amines (eg diethylamine), probably therefore at the encounter limit. However, nitrosyl thiocyanate is somewhat less reactive, as is generally the case.

$$NH_4NO_2 = N_2 + 2H_2O \qquad (4.52)$$

$$\left.\begin{array}{l} NH_3 + XNO \longrightarrow H_3\overset{+}{N}NO + X^- \\[6pt] H_3\overset{+}{N}NO \longrightarrow HN{=}N{-}OH \xrightarrow{\ H^+\ } HN_2^+ + H_2O \\[6pt] HN_2^+ \longrightarrow N_2 + H^+ \end{array}\right\} \quad (4.53)$$

4.5.2 *Hydrazoic acid*

It has long been known that azide ion in acid solution reacts rapidly with nitrous acid to give nitrous oxide and nitrogen probably via the intermediate formation of nitrosyl azide (see scheme (4.54)). Nitrosyl azide has been isolated at low temperatures,[98] but is very unstable. Its formation in the nitrosation reaction is believed to be rate-limiting. Experiments with ^{15}N also point to the intermediacy of nitrosyl azide.[99]

$$HNO_2 + HN_3 \xrightarrow{\text{slow}} N_3NO \xrightarrow{\text{fast}} N_2 + N_2O \qquad (4.54)$$

The high reactivity of the azide/hydrazoic acid system to nitrosation has made sodium azide in acid solution a particularly good nitrous acid scavenger. Indeed it is one of the most reactive systems studied.[33]

Early kinetic studies using acetate buffer solutions produced a rate law which was zero-order in azide,[100] and which was interpreted in terms of rate-limiting formation of the nitrosonium ion, but this was later shown to be a result of rate-limiting formation of nitrosyl acetate.[101] In a series of papers on the kinetics of azide nitrosation, Stedman[102] has shown that a number of reaction pathways exist, each of which can be made dominant under a given set of experimental conditions. Thus both azide ion and hydrazoic acid are reactive towards nitrous acidium ion (or nitrosonium ion) with the anion being the more reactive as expected. At low acidities the nitrosating agent is dinitrogen trioxide and generally its rate of formation is rate-limiting (see schemes (4.55) and (4.56)). All the indications are that the reaction of the azide ion is encounter-controlled. There is a remarkable similarity between the third-order rate constants k in, rate $= k[HNO_2][H^+][X^-]$ for a large number of different anions X^-, including azide ion.[103]

$$2HNO_2 \underset{\text{fast}}{\overset{\text{slow}}{\rightleftharpoons}} N_2O_3 \xrightarrow{N_3^-} N_3NO \xrightarrow{\text{fast}} N_2 + N_2O \quad (4.55)$$

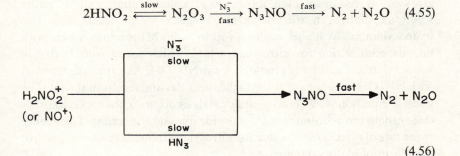

$$(4.56)$$

Catalysis by acetate, chloride, bromide and thiocyanate ions, reveals that the corresponding nitrosyl compounds are also reactive towards the azide/hydrazoic acid system. At low acidity the azide anion is probably the reactive form, but the formation of the nitrosyl species XNO is generally rate-limiting. At higher acidities reaction is first-order in $[HN_3]$ and is acid catalysed,[33] so that reaction probably occurs via hydrazoic acid.

Nitrosamines react with azide ion also, again probably giving nitrogen and nitrous oxide via nitrosyl azide. This has been established kinetically for the reaction of *N*-acetyl-*N'*-nitrosotryptophan,[104] which is sufficiently reactive to react at the low acidities necessary to generate azide ion in reasonable concentration (see scheme (4.57)).

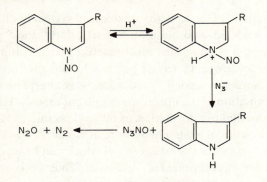

$$\text{(4.57)}$$

4.5.3 Hydroxylamines

Hydroxylamine itself reacts readily with nitrous acid in acid solution at room temperature to give nitrous oxide as shown in equation (4.58). Kinetic studies[43,105] have revealed that, in common with a number of nitrosations, several pathways exist, often taking place concurrently, the relative contribution of each depending on the experimental conditions. At low acidities, at relatively high nitrous acid concentration and in buffer solutions, dinitrogen trioxide is the reagent, which reacts with free hydroxylamine. At higher acidities (up to ~ 0.1 M) reaction occurs with nitrous acidium ion (or nitrosonium ion), which reacts with both free hydroxylamine, and its protonated form $\overset{+}{N}H_3OH$. As expected, reaction with $\overset{+}{N}H_3OH$ is dominant at the higher acidities and results in the rate law given in equation (4.59). Nucleophilic catalysis occurs in the expected order thiocyanate ion > bromide ion > chloride ion and the acidity dependence of the rate of reaction shows that the nitrosyl compounds react with the free base form of hydroxylamine.

$$NH_2OH + HNO_2 = N_2O + 2H_2O \qquad (4.58)$$

$$\text{Rate} = k[HNO_2][\overset{+}{N}H_3OH][H^+] \qquad (4.59)$$

O-Alkylhydroxylamines react similarly to give nitrous oxide and the alcohol (equation (4.60)). Experiments using $H^{15}NO_2$[106] and kinetic measurements[46] are consistent with a mechanism of direct *N*-nitrosation as outlined in scheme (4.61). Reaction involving hydroxylamine probably occurs in the same way, but detailed kinetic measurements of the reaction when $\overset{+}{N}H_3OH$ is the reactive species are not consistent with *N*-nitrosation, and it has been suggested[44] that *O*-nitrosation occurs, which is followed by *O*- to *N*-rearrangement, outlined in scheme (4.62).

$$NH_2OR + HNO_2 = ROH + N_2O + H_2O \qquad (4.60)$$

$$NH_2OR + HNO_2 \xrightarrow{\ H^+\ } ONNHOR \longrightarrow HON{=}NOR \quad (4.61)$$
$$\downarrow$$
$$N_2O + ROH$$

$$\left.\begin{array}{l} H_2NO_2^+(\text{or } NO^+) + \overset{+}{N}H_3OH \longrightarrow \overset{+}{N}H_3ONO + H_2O + H^+ \\[2ex] \overset{+}{N}H_3ONO \longrightarrow ON\overset{+}{N}H_2OH \xrightarrow{\text{fast}} N_2O + H_2O + H^+ \end{array}\right\} \quad (4.62)$$

4.5.4 Hydrazines

Hydrazine itself reacts rapidly with nitrous acid in acid solution to give ammonia and hydrazoic acid (scheme (4.63)) in ratios which depend upon the acidity of the medium. It is thought that both products arise from a common intermediate (*N*-nitrosohydrazine). With excess nitrous acid the hydrazoic acid product reacts further. Hydrazine is very reactive towards nitrosation, and has been much used as a scavenger for nitrous acid, both in laboratory experiments[33] and on the industrial scale in the Pyrex process[107] in nuclear reprocessing, specifically in the separation of uranium from plutonium, where it is used to remove nitrous acid, which would otherwise reoxidise plutonium(III) to plutonium(IV). An early literature observation[108] noted that hydrazine was much more effective than was urea in the elimination of the nitrosation pathway encountered in the nitration of phenols in concentrated nitric acid.

$$NH_2NH_2 + HNO_2 \xrightarrow{\ H^+\ } \underset{\underset{NO}{|}}{NHNH_2} \longrightarrow HO{-}N{=}N{-}NH_2$$

$$NH_3 + N_2O \qquad HN_3 + H_2O \quad (4.63)$$

With a pK_a of *ca* 8.0, hydrazine exists almost totally in the monocation form, the hydrazinium ion $N_2H_5^+$, in the acid solutions generally used in nitrosation, and kinetic measurements[37,42] confirm that the reactive species is indeed hydrazinium ion, and the reaction follows the rate equation (4.64). The reaction rate is at, or close to, the encounter limit (which may be smaller here for the reaction of two positively charged species than for the reaction of a neutral substrate). Catalysis by chloride ion, bromide ion, and thiocyanate ion again occurs and the reaction involves attack of hydrazinium ion by the corresponding nitrosyl species.

$$\text{Rate} = k[N_2H_5^+][HNO_2][H^+] \tag{4.64}$$

The hydrazinium ion is intrinsically less reactive than is hydroxylamine and other basic amines, but because of protonation equilibria, where the basic species are converted largely to their unreactive forms, the nett result is that the overall reactivity of hydrazine is much greater than for many other species. Since it is not converted to the doubly protonated species until the acid is very concentrated, hydrazine or hydrazinium ion remains one of the best nitrous acid scavengers even in high acidity.

Both alkyl- and aryl-hydrazines behave similarly kinetically and the monocations are the reactive species.[42] In these cases, however, the initial products (1-alkyl(or aryl)-1-nitrosohydrazines) break down rather differently, arylhydrazines giving aryl azides (which is a convenient synthesis of aryl azides) and alkylhydrazines a complex mixture of products (see scheme (4.65)). In the presence of excess nitrous acid, diazonium ions are obtained from phenylhydrazines, probably via a dinitroso species.

$$RN\overset{+}{H}N\overset{+}{H}_3 + H_2NO_2 \,(\text{or NO}^+) \xrightarrow{\text{slow}} RN(NO)NH_2 \longrightarrow RN_3 \quad (R = \text{aryl})$$

$$\tag{4.65}$$

$$RNH_2 + N_2O \qquad RN_3 \qquad RNHOH$$

$$(R = \text{alkyl})$$

4.5.5 *Sulphamic acid*

Sulphamic acid is another substrate much used as a nitrous acid scavenger which reacts rapidly with nitrous acid to give nitrogen and bisulphate ion (equation (4.66)). The reaction was examined kinetically by Hughes[36] at $0\,^{\circ}C$ and more recently by others[33] at $25\,^{\circ}C$. Acid catalysis occurs up to *ca* 0.2 M acid and the results are explained in terms of a reaction via the sulphamate ion form $NH_2SO_3^-$. There is no catalysis by halide ion or thiocyanate ion and there is a kinetic isotope effect of *ca* 2.7 after allowing for the effect on the protonation of nitrous acid. The results fit well into the pattern expected for an amine with a very powerful electron-withdrawing group adjacent to it, ie sulphamic acid behaves like an amide. Reaction occurs by a rapid reversible *N*-nitrosation followed by a rate-limiting proton transfer (see scheme (4.67)). No information can be deduced about the nature of the attacking nitrosating agent. The possibility exists of *O*-nitrosation followed by a rearrangement, but this has not been investigated.

$$NH_2SO_3H + HNO_2 = N_2 + HSO_4^- + H_2O + H^+ \qquad (4.66)$$

$$NH_2SO_3^- + XNO \rightleftharpoons \overset{+}{\underset{\underset{NO}{|}}{N}}H_2SO_3^- + X^- \qquad (4.67)$$

$$\downarrow \text{slow}$$

$$N_2 + HSO_4^- \longleftarrow \underset{\underset{NO}{|}}{N}HSO_3^-$$

References

1. R. Piria, *Ann. Chem. Phys.*, 1848, **22**, 160; *Annalen*, 1848, **68**, 343, and an earlier report in 1846.
2. A.W. Hofmann, *Annalen*, 1850, **75**, 356.
3. J.H. Ridd, *Quart. Rev.*, 1961, **15**, 418.
4. D.L.H. Williams, *Adv. Phys. Org. Chem.*, 1983, **19**, 381.
5. E. Kalatzis & J.H. Ridd, *J. Chem. Soc. (B)*, 1966, 529.
6. E. Muller & H. Haiss, *Chem. Ber.*, 1963, **96**, 570.
7. R.N. Butler, *Chem. Rev.*, 1975, **75**, 241.
8. H. Zollinger, *Diazo and Azo Chemistry*, Interscience, New York, 1961; H. Zollinger, *Acc. Chem. Res.*, 1973, **6**, 355; *The Chemistry of Diazonium and Diazo Groups*, Parts 1 and 2, Ed. S. Patai, Interscience, New York, 1978.
9. E.S. Lewis & R.E. Holliday, *J. Am. Chem. Soc.*, 1969, **91**, 426, and earlier papers.
10. H. Zollinger, *Angew. Chem. (Int. Edn)*, 1978, **17**, 141.
11. V. Sterba in *The Chemistry of Diazonium and Diazo Groups*, Ed. S. Patai, Interscience, New York, 1978, Ch. 2, pp. 71–93.
12. K.H. Saunders, *The Aromatic Diazo Compounds*, 2nd edn, Edward Arnold, London, 1949.
13. D.S. Wulfman in *The Chemistry of Diazonium and Diazo Groups*, Ed. S. Patai, Interscience, New York, 1978, Ch. 8, pp. 247–339.
14. J.R. Mohrig, K. Keegstra, A. Maverick, R. Roberts & S. Wells, *J. Chem. Soc., Chem. Commun.*, 1974, 780.
15. D. Semenow, C.H. Shih & W.G. Young, *J. Am. Chem. Soc.*, 1958, **80**, 5472.
16. R. Huisgen & C. Rüchardt, *Justus Liebigs Ann. Chem.*, 1956, **601**, 1, 21; A. Streitweiser & W.D. Schaeffer, *J. Am. Chem. Soc.*, 1957, **79**, 2888; A. Streitweiser, *J. Org. Chem.*, 1957, **22**, 861.

17. W. Kirmse & G. Voigt, *J. Am. Chem. Soc.*, 1974, **96**, 7598.
18. C.J. Collins, M.M. Staum & B.M. Benjamin, *J. Org. Chem.*, 1962, **27**, 3525, and earlier papers quoted therein.
19. C.J. Collins, B.M. Benjamin, V.F. Raaen, I.T. Glover & M.D. Eckart, *Justus Liebigs Ann. Chem.*, 1976, **739**, 7, and earlier papers.
20. E.H. White & D.J. Woodcock in *The Chemistry of the Amino Group*, Ed. S. Patai, Interscience, New York, 1968, Ch. 8.
21. M. Silver, *J. Am. Chem. Soc.*, 1961, **83**, 3482.
22. R.A. Moss, *J. Org. Chem.*, 1966, **31**, 1032.
23. H. Söll in *Houben-Weyl Methoden der organischen Chemie*, Ed. E. Müller, George-Thieme, Stuttgart, XI/2, 1958, pp. 131–181.
24. C.J. Collins, *Acc. Chem. Res.*, 1971, **4**, 315.
25. W. Kirmse, *Angew. Chem.* (*Int. Edn*), 1976, **15**, 251.
26. C.J. Collins, *Chem. Soc. Rev.*, 1975, **4**, 251.
27. J. Bakke, *Acta Chem. Scand.*, 1967, **21**, 1007; 1968, **22**, 1833; 1971, **25**, 859.
28. L. Friedman in *Carbonium Ions*, Vol. 2, Eds. G.A. Olah & P. von R. Schleyer, Interscience, New York, 1970, p. 655.
29. R. Weiss & K.G. Wagner, *Chem. Ber.*, 1984, **117**, 1973.
30. (*a*) E.D. Hughes, C.K. Ingold & J.H. Ridd, *J. Chem. Soc.*, 1958, 65; (*b*) E.D. Hughes & J.H. Ridd, *ibid.*, p. 70; (*c*) E.D. Hughes, C.K. Ingold & J.H. Ridd, *ibid.*, p. 77; (*d*) E.D. Hughes & J.H. Ridd, *ibid.*, p. 82; (*e*) E.D. Hughes, C.K. Ingold & J.H. Ridd, *ibid.*, p. 88.
31. G.Y. Markovits, S.E. Schwartz & L. Newman, *Inorg. Chem.*, 1981, **20**, 445.
32. H. Schmid, *Chem. Ztg.*, 1962, **86**, 811; H. Schmid & G. Muhr, *Monatsh. Chem.*, 1962, **93**, 102, and earlier papers in the series.
33. J. Fitzpatrick, T.A. Meyer, M.E. O'Neill & D.L.H. Williams, *J. Chem. Soc., Perkin Trans. 2*, 1984, 927.
34. L.F. Larkworthy, *J. Chem. Soc.*, 1959, 3304.
35. J.C.M. Li & D.M. Ritter, *J. Am. Chem. Soc.*, 1953, **75**, 3024.
36. M.N. Hughes, *J. Chem. Soc.* (*A*), 1967, 902.
37. G. Stedman & N. Uysal, *J. Chem. Soc., Perkin Trans. 2*, 1977, 667.
38. J.H. Ridd, *Adv. Phys. Org. Chem.*, 1978, **16**, 1.
39. B.C. Challis & J.H. Ridd, *J. Chem. Soc.*, 1962, 5208.
40. E.C.R. de Fabrizio, E. Kalatzis & J.H. Ridd, *J. Chem. Soc.* (*B*), 1966, 533.
41. J.T. Thompson & D.L.H. Williams, *J. Chem. Soc., Perkin Trans. 2*, 1977, 1932.
42. J.R. Perrott, G. Stedman & N. Uysal, *J. Chem. Soc., Dalton Trans.*, 1976, 2058; J.R. Perrott, G. Stedman & N. Uysal, *J. Chem. Soc., Perkin Trans. 2*, 1977, 274.
43. M.N. Hughes & G. Stedman, *J. Chem. Soc.*, 1963, 2824.
44. M.N. Hughes, G. Stedman & T.D.B. Morgan, *J. Chem. Soc.* (*B*), 1968, 344.
45. K. Singer & P.A. Vamplew, *J. Chem. Soc.*, 1956, 3971; N.S. Bayliss & D.W. Watts, *Austral. J. Chem.*, 1956, **9**, 319; N.S. Bayliss, R. Dingle & D.W. Watts, *Austral. J. Chem.*, 1963, **16**, 933.
46. B.C. Challis & J.H. Ridd, *Proc. Chem. Soc.*, 1960, 245.
47. C.G. Swain, J.T. McKnight, M.M. Labes & V.P. Kreiter, *J. Am. Chem. Soc.*, 1954, **76**, 4243.

48. L. Blangey, *Helv. Chim. Acta*, 1938, **21**, 1579.
49. H. Schmid, *Monatsh. Chem.*, 1954, **85**, 424.
50. T. Bryant, D.L.H. Williams, M.H.H. Ali & G. Stedman, *J. Chem. Soc., Perkin Trans. 2*, 1986, 193.
51. L.R. Dix & D.L.H. Williams, *J. Chem. Res. (S)*, 1984, 97.
52. E.D. Hughes & J.H. Ridd, *J. Chem. Soc.*, 1958, 82.
53. M.R. Crampton, J.T. Thompson & D.L.H. Williams, *J. Chem. Soc., Perkin Trans. 2*, 1979, 18.
54. A. Woppmann, *Monatsh. Chem.*, 1980, **111**, 1125.
55. B.C. Challis & D.E.G. Shuker, *J. Chem. Soc., Perkin Trans. 2*, 1979, 1020.
56. H. Iwamoto, T. Sonoda & H. Kobayashi, *Tetrahedron Lett.*, 1983, **24**, 4703.
57. S.E. Aldred & D.L.H. Williams, *J. Chem. Soc., Perkin Trans. 2*, 1981, 1021.
58. A. Woppmann & H. Sofer, *Monatsh. Chem.*, 1972, **103**, 163.
59. E. Kalatzis & C. Mastrokalos, *J. Chem. Soc., Perkin Trans. 2*, 1983, 53, and earlier papers in the series.
60. J. Casado, A. Castro, J.R. Leis, M.A. Lopez Quintella & M. Mosquera, *Monatsh. Chem.*, 1983, **114**, 639.
61. (*a*) S.S. Mirvish, *Toxicol. Appl. Pharmacol.*, 1975, **31**, 325; (*b*) T.Y. Fan & S.R. Tannenbaum, *J. Agric. Food Chem.*, 1973, **21**, 237.
62. A. Castro, J.R. Leis & M.E. Pena, *J. Chem. Res. (S)*, 1986, 216.
63. J.D. Okun & M.C. Archer, *J. Natl. Cancer Inst.*, 1977, **58**, 409.
64. K. Heyns, H. Röper, S. Röper & B. Meyer, *Angew. Chem.*, 1979, **91**, 940.
65. S.S. Mirvish, J. Sams, T.Y. Fan & S.R. Tannenbaum, *J. Natl. Cancer Inst.*, 1973, **51**, 1833.
66. J. Casado, A. Castro, J.R. Leis, M. Mosquera & M.E. Pena, *J. Chem. Soc., Perkin Trans. 2*, 1985, 1859.
67. A. Castro, E. Iglesias, J.R. Leis, M.E. Pena, J.V. Tato & D.L.H. Williams, *J. Chem. Soc., Perkin Trans. 2*, 1986, 1165.
68. B. Guether, *Arch. Pharm.*, 1864, **2**, 123, 200.
69. G.E. Hein, *J. Chem. Educ.*, 1963, **40**, 181.
70. P.A.S. Smith & R.N. Loeppky, *J. Am. Chem. Soc.*, 1967, **89**, 1147.
71. B.G. Gowenlock, R.J. Hutchison, J. Little & J. Pfab, *J. Chem. Soc., Perkin Trans. 2*, 1979, 1110.
72. W. Lijinsky, L. Keefer, E. Conrad & R. van de Bogart, *J. Natl. Cancer Inst.*, 1977, **49**, 1239.
73. R.N. Loeppky & W. Tomasik, *J. Org. Chem.*, 1983, **48**, 2751.
74. L.K. Keefer in *N-Nitrosamines*, Ed. J-P. Anselme, ACS Symposium Series 101, American Chemical Society, Washington D.C., 1979, pp. 94–8.
75. S.S. Hecht, C.B. Chen, R.M. Ornaf, E. Jacobs, J.D. Adams & D. Hoffman, *J. Org. Chem.*, 1978, **43**, 72.
76. R.N. Loeppky, J.R. Outram, W. Tomasik & J.M. Faulconer, *Tetrahedron Lett.*, 1983, **24**, 4271.
77. N. Sperber, D. Papa & E. Schwenk, *J. Am. Chem. Soc.*, 1948, **70**, 3091; Z. Kricsfalussy and A. Bruylants, *Bull. Soc. Chim. Belges.*, 1964, **73**, 96.
78. G.A. Olah & J.A. Olah, *J. Org. Chem.*, 1965, **30**, 2386.
79. E.H. White, *J. Am. Chem. Soc.*, 1955, **77**, 6008.

80. Z. Kricsfalussy & A. Bruylants, *Bull. Soc. Chim. Belges.*, 1967, **76**, 168 and earlier papers.
81. B.C. Challis & J.A. Challis in *The Chemistry of Amides*, Ed. J. Zabicky, Interscience, New York, 1970, pp. 780–4.
82. (*a*) M. Yamamoto, T. Yamada & A. Tanimura, *J. Food Hyg. Soc. Jpn.*, 1976, **17**, 363; (*b*) C.N. Berry & B.C. Challis, *J. Chem. Soc., Perkin Trans. 2*, 1974, 1638.
83. G. Hallett & D.L.H. Williams, *J. Chem. Soc., Perkin Trans. 2*, 1980, 1372.
84. J.K. Snyder & L.M. Stock, *J. Org. Chem.*, 1980, **45**, 886.
85. J. Casado, A. Castro, M. Mosquera, M.F. Prieto & J.V. Tato, *Ber. Bunsenges. Phys. Chem.*, 1983, **87**, 1211; J. Casado, A. Castro, J.R. Leis, M. Mosquera & M.E. Pena, *Monatsh. Chem.*, 1984, **115**, 1047.
86. B.C. Challis & S.P. Jones, *J. Chem. Soc., Perkin Trans. 2*, 1975, 153.
87. A. Castro, E. Iglesias, J.R. Leis, M.E. Pena & J.V. Tato, *J. Chem. Soc., Perkin Trans. 2*, 1986, 1725.
88. R.B. Horner & C.D. Johnston in *The Chemistry of Amides*, Ed. J. Zabicky, Interscience, New York, 1970, p. 188.
89. B.C. Challis, J.N. Iley & H.S. Rzepa, *J. Chem. Soc., Perkin Trans. 2*, 1983, 1037.
90. D.D. van Slyke, *J. Biol. Chem.*, 1929, **83**, 425 and earlier papers.
91. B.C. Challis, J.R. Milligan & R.C. Mitchell, *J. Chem. Soc., Chem. Commun.*, 1984, 1050; J.R.A. Pollock, in *N-Nitroso Compounds: Occurrence and Biological Effects*, Eds. H. Bartsch, I.K. O'Neill, M. Castegnaro and M. Okada, IARC Scientific Publication No. 41, Lyon, 1982, p. 81.
92. J. Garcia, J. Gonzalez, R. Segura & J. Vilarrasa, *Tetrahedron*, 1984, **40**, 3121.
93. D.L.H. Williams, *J. Chem. Soc., Perkin Trans. 2*, 1976, 1838.
94. J.H. Dusenbury & R.E. Powell, *J. Am. Chem. Soc.*, 1951, **73**, 3266; G.J. Ewing & N. Bauer, *J. Phys. Chem.*, 1958, **62**, 1449.
95. G.A. Olah, R. Herges, J.D. Felberg & G.K.S. Prakash, *J. Am. Chem. Soc.*, 1985, **107**, 5282.
96. H. Schmid & R. Pfeifer, *Monatsh. Chem.*, 1953, **84**, 829; H. Schmid, *Monatsh. Chem.*, 1954, **85**, 424.
97. T. Bryant & D.L.H. Williams, *J. Chem. Soc., Perkin Trans. 2*, 1987, in the press.
98. H.W. Lucien, *J. Am. Chem. Soc.*, 1958, **80**, 4458.
99. K. Clusius & E. Effenberger, *Helv. Chim. Acta*, 1955, **38**, 1834, 1843.
100. F. Seel & E. Schwaebel, *Z. Anorg. Allg. Chem.*, 1953, **274**, 169.
101. C.A. Bunton, D.R. Llewellyn & G. Stedman, *Chem. Soc. Spec. Publ.*, 1957, **10**, 113.
102. G. Stedman, *J. Chem. Soc.*, 1959, 2943, 2949, 3466; *ibid.*, 1960, 1702.
103. G. Stedman, *Adv. Inorg. Chem. Radiochem.*, 1979, **22**, 113.
104. T.A. Meyer, D.L.H. Williams, R. Bonnett & S.L. Ooi, *J. Chem. Soc., Perkin Trans. 2*, 1982, 1383.
105. C. Döring & H. Gehlen, *Z. Anorg. Allg. Chem.*, 1961, **312**, 32.
106. J.E. Leffler & A.A. Bothner-By, *J. Am. Chem. Soc.*, 1951, **73**, 5473.
107. E.K. Dukes, *J. Am. Chem. Soc.*, 1960, **82**, 9; P. Biddle & J.H. Miles, *J. Inorg. Nucl. Chem.*, 1968, **30**, 1929.
108. F. Arnall, *J. Chem. Soc.*, 1923, **123**, 3111.

5

N-Nitrosamines

This chapter discusses only the reactions of *N*-nitrosamines which can be regarded as nitrosation reactions, ie reactions which involve the transfer of the nitroso group to another species, or to another site within the same molecule (a rearrangement). These reactions can occur heterolytically or homolytically and are discussed under these headings. In addition a section at the end of the chapter (section 5.4) is devoted to the carcinogenic behaviour of nitrosamines, although this does not involve nitrosation; it is included because of the possible hazards involved in working with nitrosamines or by their *in vivo* formation from food products and water supplies. Nitrosamines have been known since the early days of organic chemistry, and undergo many reactions which will not be covered here. Aspects of the structure, formation and reactions of nitrosamines, have been covered elsewhere in the literature,[1] and there are many monographs[2] and review articles devoted to the carcinogenicity of nitrosamines. In addition the International Agency for Research on Cancer (IARC) has published the proceedings of a number of conferences specifically dealing with the occurrence, formation, analysis and biological effects of *N*-nitroso compounds, principally *N*-nitrosamines and *N*-nitrosamides.[3]

5.1. Rearrangement of aromatic *N*-nitrosamines (the Fischer–Hepp rearrangement)

One of the earliest known reactions of nitrosamines concerns aromatic systems in which the nitroso group rearranges from the amino nitrogen atom to the aromatic ring, usually to the 4-position. This reaction can be regarded as an internal *N*- to *C*-nitrosation; it occurs in a range of solvents, and is acid-catalysed, usually, but not necessarily, by hydrogen chloride. It is general for a range of R groups (amino substituents) and of X groups (aromatic ring substituents) other than for 4-X substituents, as outlined in equation (5.1). The reaction formally resembles a number of other acid-catalysed rearrangements of *N*-substituted aniline derivatives, such as the benzidine rearrangement of hydrazobenzenes, the Orton rearrangement of *N*-chloroanilides, the nitramine rearrangement of *N*-nitroanilines, the Bamberger rearrangement of phenylhydroxylamines, and a number of other reactions. All of these reactions, which have been discussed in detail elsewhere,[4,5] have found some use in synthetic organic chemistry, and have also posed many interesting mechanistic problems. The mechanism of the benzidine rearrangement for example has been the subject of a vast amount of work. It transpires that a range of different mechanisms exists within the group of acid-catalysed rearrangements of *N*-substituted aniline derivatives, so that any similarities are, in fact, only superficial.

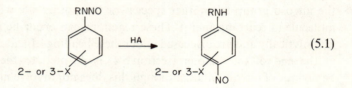

$$\text{(5.1)}$$

The rearrangement of aromatic *N*-nitrosamines was discovered in 1886 by Fischer and Hepp,[6] after whom the reaction is often known. They established suitable experimental conditions for the maximum yields of the *C*-nitroso isomer, which are still in use today, at least in laboratory preparations. Generally the nitrosamine in dry ether or ethanol is added to a solution of hydrogen chloride in ethanol at room temperature. Reaction is fairly rapid and the product is precipitated as the hydrochloride. The range of R groups used in the early work includes $R = CH_3$, C_2H_5, C_6H_5, 4-ClC_6H_4, $(CH_3)_2CH_2CH_2$, and $(CH_3)_2CH_2$;[6,7] similar reactions occur in the naphthylamine series (see equations (5.2) and (5.3)).[8,9] Interestingly there is no report of a 2-substituted nitroso amine product (other than in the 2-naphthyl series), and indeed 2-nitroso aniline derivatives have only very rarely been characterised. The 4-isomer is also the major (or only) product in some other rearrangement reactions of this type, eg the rearrangement of *N*-alkyl anilines,[10] the Bamberger rearrangement of phenylhydroxylamines[11] etc. Further, the nitrosation of phenols and phenyl ethers also results predominantly in the formation of 4-nitroso products.[12]

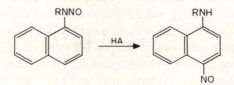

$$\text{(5.2)}$$

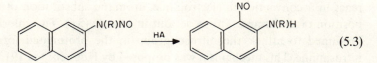

$$(5.3)$$

The product of denitrosation – the secondary amine – is often formed along with the rearrangement product, and occasionally dominates. This is particularly true when the aromatic ring contains 2- or 3-substituted electron-attracting substituents.[13]

Two somewhat unusual examples of the rearrangement are shown in equations (5.4) and (5.5). The first involves the rearrangement of two nitroso groups in a dinitroso derivative of 3-phenylenediamine.[14] The second probably involves the Fischer–Hepp rearrangement in an initially formed *N*-nitroso ruthenium derivative, in the reaction of a ruthenium nitrosyl cation with *N*-methylaniline, to give finally the 4-nitroso arene complex.[15]

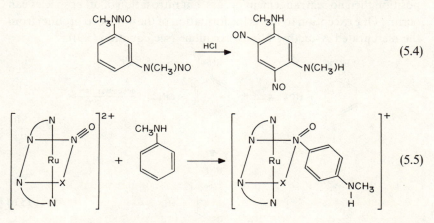

$$(5.4)$$

$$(5.5)$$

The Fischer–Hepp rearrangement is carried out on the industrial scale to prepare 4-nitrosodiphenylamine (from *N*-nitrosodiphenylamine) which is used to make Diafen FP (*N-iso*propyl-*N'*-phenyl-*para*-phenylenediamine) a rubber antioxidant. There is a large patent literature describing modifications to the rearrangement reaction, a typical procedure is to treat the nitrosamine in toluene solution with 30% hydrogen chloride in methanol in a cascade of interconnected reactors at room temperature.[16] Rearrangement also occurs in good yield in aprotic solvents and with Lewis acid catalysts such as aluminium chloride.[17]

For over fifty years the reaction mechanism has been expressed in terms of a reversible acid-catalysed, chloride ion promoted denitrosation of the nitrosamine to give the secondary amine and nitrosyl chloride, which then react in a conventional, electrophilic, aromatic substitution of the *para*-position of the amine. This is set out in scheme (5.6). The chloride ion is presumed to attack the nitroso group in the protonated form of the nitrosamine. This mechanism was proposed by Houben[18] in 1913, and the evidence in its favour derives mainly from three sets of observations: (*a*) that the yields of rearrangement products are noticeably greater when hydrogen chloride is the acid catalyst, implying that chloride ion is somehow directly involved; (*b*) that a number of nitrosations of species such as *N,N*-dimethylaniline or reactive alkenes, added to the reaction mixture, can be achieved. Thus the free nitrosating reagent (nitrosyl chloride) is being trapped, giving 4-nitroso-*N,N*-dimethylaniline and the nitroso chloro adduct of the alkene (equations (5.7) and (5.8));[9] (*c*) the reaction of 3-nitro-*N*-methyl-*N*-nitrosoaniline in the presence of urea gives only the product of denitrosation.[19] When the nitrosamine is already substituted in the 4-position then no rearrangement occurs, but nitrosation of other species can occur.[9] One exception to this is the formation of the 4-nitroso product from the mercurated *N*-methyl-*N*-nitrosoaniline (see equation (5.9)).[20]

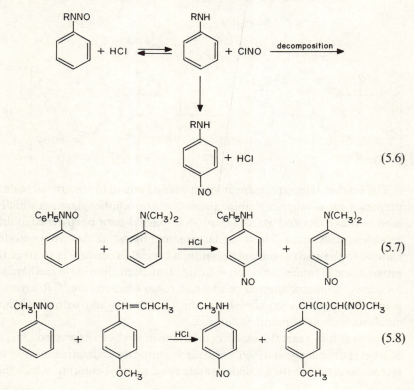

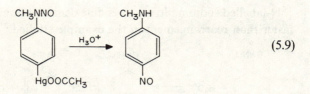

$$(5.9)$$

Whilst all of the above evidence is consistent with the mechanism outlined in scheme (5.6), it by no means specifically demands such a mechanism. The fact that an intermediate can be trapped does not necessarily mean that it is an intermediate on the pathway leading to rearrangement. Further, point (c) above was totally inconclusive, since it was not shown that rearrangement occurred in the absence of urea. In fact, later work[21] has shown that this nitrosamine gives only the denitrosation product. Nevertheless this mechanism has been generally accepted, and has been widely presented in text books and review articles, although one or two authors[22] have recognised that the evidence was inconclusive and that a full mechanistic investigation is overdue.

It was noted by two independent groups[23,24] that rearrangement could occur even in the presence of quite high concentrations of 'nitrite traps' such as urea or sulphamic acid. Kinetic measurements on the rearrangement of *N*-nitrosodiphenylamine[25] in methanol showed that although the reaction was acid-catalysed it was not subject to chloride ion catalysis. This contrasts with the situation found for the rearrangement of *N*-chloroanilides[26] where the rate law includes a term in $[H^+]$ and also in $[Cl^-]$, and which is readily interpreted in terms of an intermolecular mechanism similar to scheme (5.6) with molecular chlorine as the reactive intermediate. An alternative mechanism was suggested[27] in 1972 in which rearrangement takes place *intramolecularly* and in parallel with reversible denitrosation. This is outlined in scheme (5.10). This scheme explains equally well all the early product analyses, including the possibility of effecting nitrosation of added nucleophiles such as amines or alkenes. In many cases, particularly the rearrangements of other *N*-substituted anilines, it is possible to make a clear distinction between an intra and an intermolecular mechanism using isotope experiments,[28] noting the exchange or otherwise of label between the migrating species and a similar one added to the reaction mixture. In this way the intramolecularity of the benzidine rearrangement and the intermolecularity of the Bamberger rearrangement of phenylhydroxylamines, were confirmed. However in this case, such experiments are of no value since the reactant nitrosamine exchanges its nitroso group rapidly with nitrous acid in solution, added as

[15]N-labelled sodium nitrite; reversible denitrosation is indeed significantly
faster than rearrangement in the example studied.[27]

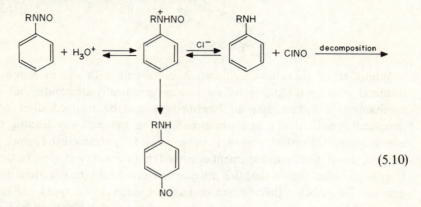

(5.10)

A clear distinction between the two possible mechanisms has, however,
been made on the basis of the ratio of the products of rearrangement and
denitrosation when the reaction is carried out in the presence of added
nitrite traps such as sulphamic acid. In the absence of such added species,
the relative yields of rearrangement and denitrosation are governed by a
number of factors including the structure of the nitrosamine, the nature and
concentration of the nucleophile, and the decomposition rate of nitrosyl
chloride (or equivalent species) in solution. If the intermolecular mechanism operates (scheme (5.6)) then it is to be expected that the yield of
rearrangement product should decrease, eventually to zero as the concentration of a nitrite trap (sulphamic acid, hydrazine, hydrazoic acid, urea,
hydroxylamine etc.) is increased, since there is now a direct competition

Table 5.1. *Rearrangement yield and overall rate constant for reaction of
N-methyl-N-nitrosoaniline in sulphuric acid (2.75 M) in the presence of
various nitrite traps*

Nitrite trap	% Rearrangement	$10^4 k/s^{-1}$
Hydrazoic acid (6.5×10^{-4} M)	21	0.65
Hydrazoic acid (16.3×10^{-4} M)	21	0.67
Sulphamic acid (3.1×10^{-3} M)	21	0.65
Sulphamic acid (7.8×10^{-3} M)	22	0.64
Urea (0.10 M)	21	0.62
Hydroxylamine (1.6×10^{-3} M)	20	0.66
Hydroxylamine (2.6×10^{-3} M)	20	0.62

between the secondary amine and the nitrite trap for the free nitrosating species. For the alternative mechanism (scheme (5.10)) the yield of rearrangement product should also decrease with increasing nitrite trap concentration, but not now to zero but to a constant value, when the re-*N*-nitrosation of the secondary amine has been effectively cut out, and we now have a situation of two parallel irreversible reactions. For a given nitrosamine the limiting yield of rearrangement product will depend again on the nature of the nucleophile (chloride ion, bromide ion, thiocyanate ion etc.) and on its concentration, but should be independent of the nature and concentration of the nitrite trap once the threshold value (necessary to eliminate re-*N*-nitrosation) has been achieved. The results of such experiments in water solvent, are given in tables 5.1, 5.2 and 5.3. In table 5.1 the reactant nitrosamine is *N*-methyl-*N*-nitrosoaniline (**5.1**) and it is clear that the % rearrangement under these conditions is constant at $\sim 21\%$ for a

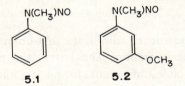

5.1   **5.2**

range of nitrite trap species and concentrations.[29] Similarly in table 5.2, where the reactant is *3*-methoxy-*N*-methyl-*N*-nitrosoaniline (**5.2**) the rearrangement yield is again constant this time at $\sim 85\%$ over a range of nitrite trap conditions.[30] These results are quite incompatible with the intermolecular mechanism, but are expected from the intramolecular mechanism outlined in scheme (5.10). The observed first-order rate constants (*k*) are also constant within tables 5.1 and 5.2, as expected from

Table 5.2. *Rearrangement yield and overall rate constant for reaction of 3-methoxy-N-methyl-N-nitrosoaniline in sulphuric acid (3.5 M) in the presence of various nitrite traps*

Nitrite trap	% Rearrangement	$10^3 k/s^{-1}$
–	95	2.94
Hydrazoic acid $(1 \times 10^{-3} \text{ M})$	84	3.10
Hydrazoic acid $(5 \times 10^{-3} \text{ M})$	85	3.22
Hydrazine $(1 \times 10^{-3} \text{ M})$	85	3.25
Hydrazine $(5 \times 10^{-3} \text{ M})$	85	3.39
Sulphamic acid $(1 \times 10^{-3} \text{ M})$	84	3.33
Sulphamic acid $(5 \times 10^{-3} \text{ M})$	84	3.48

Table 5.3. *Rearrangement yield and overall rate constant for the reaction of 3-methoxy-N-methyl-N-nitrosoaniline in sulphuric acid (3.5 M) containing hydrazoic acid (5 × 10⁻³ M) and added nucleophiles*

Nucleophile	% Rearrangement	$10^3 \, k/\text{s}^{-1}$
(Water)	80	3.27
Chloride ion (0.077 M)	73	3.51
Bromide ion (0.077 M)	16	–
Thiocyanate ion (0.077 M)	0	–
Thiourea (0.077 M)	0	–
Bromide ion (0.004 M)	65	3.86
Bromide ion (0.008 M)	56	4.14
Bromide ion (0.016 M)	36	6.6
Bromide ion (0.032 M)	29	9.4
Bromide ion (0.100 M)	11	–

scheme (5.10), and represent the sum of the first-order rate constants for rearrangement and denitrosation. The latter is dependent upon the reactivity of the nucleophile and includes the nucleophile concentration, so it is to be expected that the yield of rearrangement product would decrease, and the observed rate constant would increase as the nucleophile reactivity and the nucleophile concentration are increased. This is borne out experimentally as shown by the data in table 5.3 for the reaction of **5.2**.[30] Similar results were obtained for the reaction of **5.2** in ethanol solvent where the rearrangement yield decreased and the observed rate constant increased as the concentration of added thiourea was increased, both for sulphuric acid and hydrogen chloride acid catalysts. This is shown graphically for the experiments using hydrogen chloride in figure 5.1. It appears that in ethanol, the solvent itself can act as a nitrite trap by the formation of ethyl nitrite which is not an active reagent towards amines in acid solution. Thus the reaction can be represented by two parallel reactions one leading to denitrosation and the other to rearrangement. Then as expected the first-order rate constant k is linear with [thiourea] with an intercept which represents the rate constant for rearrangement together with that for denitrosation by the solvent and hydrogen chloride. A more detailed kinetic analysis enables the proportion of rearrangement to be calculated; these values are included on the graph in figure 5.1 and show excellent agreement with the observed values.

When significant amounts of denitrosation occur, normally it should be possible to suppress this reaction in favour of rearrangement by addition of

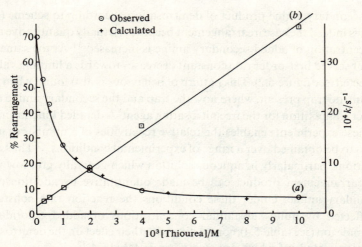

Figure 5.1 Variation of (*a*) rearrangement yield and (*b*) overall rate constant for the reaction of 3-methoxy-*N*-methyl-*N*-nitrosoaniline in hydrogen chloride–ethanol as a function of added thiourea.

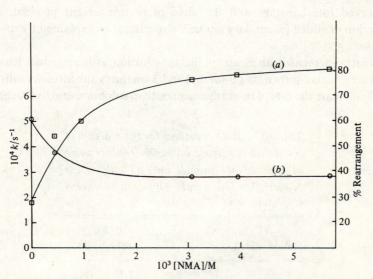

Figure 5.2 Variation of (*a*) rearrangement yield and (*b*) overall rate constant for the reaction of *N*-methyl-*N*-nitrosoaniline in aqueous sulphuric acid as a function of added *N*-methylaniline (NMA).

the secondary amine product of denitrosation according to scheme (5.10). This is indeed the case; rearrangement becomes virtually quantitative as the concentration of added secondary amine is increased.[31] At the same time the observed first-order rate constant decreases towards a limiting value as expected (see figure 5.2). This pattern of behaviour is also found when there is a nitrite trap present, where now the trap and the secondary amine are in direct competition for the free nitrosating agent. A detailed kinetic analysis of these experiments enables the relative reactivities of a number of nitrite traps to be obtained over a range of experimental conditions.[32] Thus many reactions, particularly in aqueous solution which generally give low yields of rearrangement product can be made quantitative by addition of the secondary amine. Under these conditions the reaction rate constant is unaffected by quite substantial concentrations of added chloride and bromide ion (see table 5.4), contrasting with their effect on the denitrosation reaction, details of which are also given in table 5.4.[33]

None of these observations can be explained by the intermolecular mechanism (scheme (5.6)) which must now be discarded. Equally, the experimental results are not consistent with a mechanism involving two pathways to rearrangement, one intramolecular and one intermolecular. Particularly difficult to explain in this way is the independence of the rate constant for rearrangement upon [halide ion]. The variation in both the observed rate constant and the yield of rearrangement product, as a function of added [secondary amine], also cannot be explained by scheme (5.6).

Rearrangement is an electrophilic substitution as is expected. Russian workers[34] have shown that 3-methyl and 3-methoxy substituents substantially increase the rate of rearrangement both in sulphuric and hydrochloric

Table 5.4. *Rate constants for the reaction of N-methyl-N-nitrosoaniline in the presence of chloride and bromide ion (a) with excess N-methylaniline and (b) with excess sulphamic acid*

Halide added	$10^4\,k/s^{-1}$	
	(a)	(b)
–	1.75	16.3
Sodium chloride (0.24 M)	1.79	41.7
Sodium bromide (0.10 M)	1.77	75.1

acids. Similar results were obtained later,[31] which also demonstrated a much reduced reactivity for the 3-chloro nitrosamine. As mentioned earlier, the deactivating effect of the 3-nitro substituent is so great that no rearrangement takes place at all. A 2-methyl substituent also activates slightly, probably by affecting the basicity of the nitrosamine, whereas for the 2, 6-dimethyl compound reaction is very slow indeed, being just detectable over several days.[35] This clearly is a steric effect, probably steric hindrance to solvation of the protonated nitrosamine, since the reaction leading to denitrosation is similarly affected.

Acid catalysis of rearrangement has been established quantitatively by a number of workers, and the reaction is, as expected, subject to a solvent isotope effect k_{H_2O}/k_{D_2O} of *ca* 0.3. Rate constants correlate quite well with the Hammett acidity function h_0. This suggests that reaction occurs via the protonated form of the nitrosamine and that protonation is rapid (see equation (5.11)). Protonation is written as occurring at the amino nitrogen atom, which seems to fit in better with a subsequent transfer of the nitroso group, than would be the case for oxygen protonation. At low acidities the extent of protonation is expected to be small, but at higher acidities the rate constant for rearrangement levels off, and there is a change in the uv spectrum, suggesting that protonation is extensive in this region, this enables the pK_a of N-methyl-N-nitroso-aniline to be estimated[36] as *ca* -2. The final proton transfer from the *para*-position to the solvent is subject to a primary kinetic isotope effect, k_H/k_D (for ring-deuterated substrates) values are 2.4[31] for N-methyl-N-nitrosoaniline and vary from 1.4 to 2.4 for the 3-methoxy nitrosamine as the acidity is increased from 0.8 M to 3.2 M sulphuric acid.[30] This identifies the σ complex as an intermediate (see equation (5.12)) and suggests that the final proton loss to the solvent is at least in part rate-limiting, which is not the commonly encountered situation in electrophilic aromatic substitution.

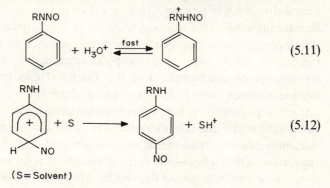

(5.11)

(5.12)

(S = Solvent)

The increase of k_H/k_D with acidity has been taken to reflect the decreasing water activity, resulting in a reduced rate of proton loss. There is evidence that this is an important factor in rearrangement at very high acidities (> 10 M sulphuric acid) where the rate constant for rearrangement decreases with acidity, denitrosation now competes more effectively and yields of rearrangement product are low.[36] This fits in with the observation made by Fischer and Hepp that in concentrated sulphuric acid a nitrosamine remains unchanged over several weeks.[5]

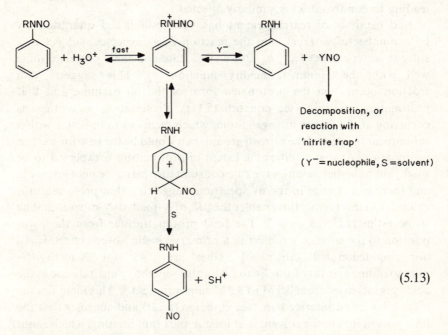

$$(5.13)$$

The collected experimental observations can be accommodated by the mechanism given in scheme 5.13. One major unanswered question is, how does the intramolecular rearrangement actually take place? Other aromatic intramolecular rearrangements have been explained in terms of a caged radical-ion mechanism, a π complex mechanism, or a polar cyclic transition-state mechanism. For the Fischer–Hepp rearrangement no definite evidence exists in favour of any of these alternatives, although an attractive possibility is that proposed initially for the rearrangement of sulphanilic acid[37] and later extended[38] as a possible general mechanism for intramolecular rearrangements of *N*-substituted aniline derivatives. The suggestion is that a bent-boat form intermediate is formed, which is ring-protonated (see scheme (5.14)), and in which the nitroso group is near enough to the 4-position to allow binding to occur in one stage without the need for a further intermediate. At this point, this must remain a matter of speculation, as is the reason why no 2-substituted rearrangement product is formed.

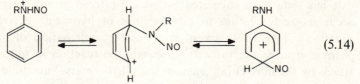

$$(5.14)$$

Before leaving the Fischer–Hepp rearrangement it must be recorded that Russian workers have published a series of papers[39] in which they maintain that the rearrangement occurs intermolecularly. Their approach is to assume such a mechanism (see scheme (5.15)) and attempt to obtain separate values independently for the rate constants k_1, k_2 and k_3, and then to compute the expected rate profile (concentration–time) which is then compared with that observed.[40] This treatment is by no means a definitive one for the intermolecular mechanism, and can be criticised on a number of grounds; *viz* the treatment is not extended to the intramolecular case and so no comparison can be made between the two mechanisms; k_3 values are obtained from the nitrosation of *N*-substituted diphenylamines $(C_6H_5)_2NR$, where it is assumed that *C*-nitrosation occurs directly, whereas the initial attack could be at nitrogen followed by intramolecular rearrangement; many of the determinations rely on an interpolation procedure of Hammett type correlations, using a small number (typically four) of points to define the line, so that possible errors are quite large. Finally it is quite impossible to reconcile many of the more recent experimental observations with the intermolecular mechanism, the crucial experiments being the constant rearrangement yield at high concentrations of nitrite trap, the absence of halide ion catalysis for rearrangement, and the increasing yield of rearrangement product (and the simultaneously decreasing rate constant), found as the secondary amine product is added.

$$(C_6H_5)_2NNO + HCl \underset{k_2}{\overset{k_1}{\rightleftharpoons}} (C_6H_5)_2NH + ClNO$$
$$\downarrow k_3$$
$$ONC_6H_4N(C_6H_5)H \qquad (5.15)$$

5.2. Other internal nitroso group rearrangements involving nitrosamines

A number of reactions are known where the nitroso group in a nitrosamine is transferred to another site within the same molecule, although none of these reactions has been subjected to the detailed mechanistic study accorded to the Fischer–Hepp rearrangement.

$$RN(NO)NHR' \longrightarrow RNHN(NO)R' \qquad (5.16)$$

It has long been suggested[41] that 1, 2-nitroso group rearrangements occur in acid solution in nitrosamines of hydrazine derivatives (see equation (5.16)). More recent work on the nitrosation of methyl and phenylhydrazine derivatives[42] discusses the reaction in terms of initial attack by the nitrosating agent at the alkyl- or aryl-substituted nitrogen atom, followed by a rearrangement to the terminally-substituted nitroso compound (see scheme (5.17)). The formation of the various products is best rationalised by reaction of the terminally-substituted nitroso compound, but initial nitrosation is believed to occur at the other nitrogen atom, since the terminal amino group is virtually completely protonated at the acidity of the experiments. This nitrosamine can in fact be isolated and in the case of the phenylhydrazine derivative, the subsequent reaction in acid solution to give phenyl azide, is also believed to involve a 1, 2-nitroso group rearrangement (see scheme (5.18)).[43]

$$RNHNH_2 \xrightarrow[H_3O^+]{HNO_2} R\overset{+}{N}H(NO)NH_2 \longrightarrow RNHNH(NO)$$
$$\longrightarrow \text{Various products} \qquad (5.17)$$

$$C_6H_5N(NO)NH_2 \xrightarrow{H_3O^+} C_6H_5NHNH(NO) \longrightarrow C_6H_5N_3 \quad (5.18)$$

A 1, 3-nitroso group rearrangement occurs when 1-nitroso-1, 2, 3, 4-tetrahydro-pyridine is treated with hydrochloric acid or with trifluoroacetic acid (see equation (5.19)),[44] when the product oxime is isolated as the salt. A possible mechanism is the formation of the carbocation by addition of a proton to the double bond, and reaction of this species as a nitrosating agent.

$$(5.19)$$

Such rearrangements can also occur in base solution, an example of a 1, 2-nitroso group shift is in the rearrangement of a *N*-nitroso cyano species to give the isomeric oxime (see equation (5.20)).[45] It is suggested that the base abstracts a proton from the cyanomethyl group and that the 1, 2-nitroso group shift occurs by nucleophilic attack by the carbanion centre at the nitroso nitrogen atom, possibly via a three-membered ring intermediate. An example of a base-catalysed 1, 3-nitroso group rearrangement is the isomerisation of *N*-nitrosohydrazines (see equation (5.21)).[46]

$$RN(NO)CH_2CN \xrightarrow{OH^-} \underset{\substack{\| \\ NOH}}{RNHCCN} \tag{5.20}$$

$$C_6H_5N(NO)N{=}CHC_6H_5 \xrightarrow{OR^-} \underset{\substack{\| \\ NOH}}{C_6H_5N{=}N{-}CC_6H_5} \tag{5.21}$$

Nitroso group rearrangements can also occur by free-radical pathways, almost certainly by intermolecular mechanisms. Three such examples are shown in equations (5.22),[47] (5.23)[48] and (5.24).[49] In each case it is believed that homolysis of the N—N bond occurs giving nitric oxide and the aminium radical $R_2NH^{+\cdot}$, which is then subjected to free-radical attack by nitric oxide at some other point in the molecule. These reactions have some similarity to the Barton reaction of alkyl nitrites (see section 6.5), as do the photolyses of nitrosamides, which result in the formation of δ-*C*-nitroso compounds isolated as dimers.

$$\underset{R'CH_2}{\overset{R}{\diagdown}}N{-}NO + H_3O^+ \xrightarrow{h\nu} \underset{\substack{| \\ R'}}{RNHC{=}NOH} \tag{5.22}$$

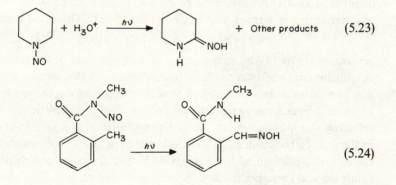

(5.23)

(5.24)

Another example of a 1, 3-nitroso group transfer believed to involve free-radical intermediates is the reaction of an alkene-substituted nitrosamine to give the iminooxime (see equation (5.25)).[50] This reaction can be brought about thermally, photochemically or by reaction with hydrogen chloride. Again it is likely that nitric oxide is formed which attacks the alkene carbon

atom. Indeed it is quite common for photoaddition reactions of nitrosamines with a variety of substrates to be accompanied by secondary
reactions which yield *C*-nitroso compounds usually stabilised in their
oxime form.

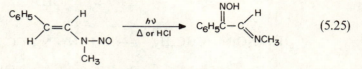

$$(5.25)$$

5.3. Nitrosamines as nitrosating agents

This section will be concerned with reactions in which the nitroso group
from a nitrosamine is transferred to some other species. The N—N bond
can undergo heterolytic or homolytic cleavage; in principle both processes
lead to a fragment capable of effecting nitrosation. The alternative
possibility is that direct transfer of the NO group can occur without the
intermediacy of a free nitrosating agent. It turns out that all three reaction
types occur, and these will be discussed in turn.

5.3.1 *N—N Heterolytic cleavage with direct transfer of the nitroso group*

Probably the earliest recognised reaction of this type is the acid-catalysed
denitrosation of a nitrosamine to give the secondary amine and nitrous acid
or a derivative of nitrous acid (see equation (5.26)). This is, of course, the

$$RR'NNO \xrightarrow{H_3O^+} RR'NH + HNO_2 \qquad (5.26)$$

reverse reaction of the preparation of nitrosamines and generally the
equilibrium lies well over to the nitrosamine side, so that denitrosation will
not be evident unless the nitrous acid is in some way removed from the
solution. Procedures which have proved useful synthetically include
refluxing the nitrosamine in alcoholic sulphuric acid with added urea,[19]
or the use of a reducing agent such as cuprous or ferrous ion which reduces
the nitrous acid formed to nitric oxide.[51] The reactions are clearly acid-
catalysed and therefore involve the protonated form of the nitrosamine as
the reactive species. There are of course two possible protonation sites in
nitrosamines, at the amino nitrogen atom and at the nitroso oxygen atom
(see **5.3** and **5.4**). There has been some uncertainty about which form

$$R_2\overset{+}{N}HNO \qquad\qquad R_2N\overset{+}{N}OH$$

5.3   **5.4**

dominates, as there was in the controversy regarding protonation of amides, but it now seems likely that the nitroso oxygen atom is the more basic site. This allows resonance stabilisation of the positive charge (see **5.5**).

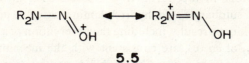

5.5

Physical evidence in favour of O-protonation comes from nmr studies[52] of nitrosamines in strong acids (such as 'magic acid'), where on protonation, the two methyl groups of nitrosodimethylamine are non-equivalent, implying that free rotation about the N—N bond is not possible. Calculations also indicate that the O-protonated form of the nitrosamine is more stable than is the N-protonated form.[53] However, it seems likely that denitrosation of nitrosamines occurs by reaction via the N-protonated form which is probably present in very low concentrations. Arguments in favour of this suggestion centre around the fact that it is easier to write a plausible mechanism for the denitrosation process via the N-protonated form, given the principle of microscopic reversibility requires that the same mechanism operates, in a reverse sequence of steps, for the N-nitrosation of amines. It is not unusual for reaction to proceed via the one (of two possible) intermediate which is present in the smaller concentration, as for example in the Bamberger rearrangement of N-phenyl-hydroxylamines.[54] The extent and rates of denitrosation are very dependent upon the presence of added (non-basic) nucleophilic species eg halide ion. The reaction has been much used in the quantitative analysis of nitrosamines using the so-called Thermal Energy Analyser TEA. A typical procedure[55] is to treat the nitrosamine with 5–10% solution of hydrogen bromide in acetic acid, and to convert the nitrosyl bromide produced to nitric oxide which is analysed quantitatively by means of a chemiluminescence detector following reaction with ozone when nitrogen dioxide in an excited state is formed. Other solvents and other nucleophiles have also been used successfully. An alternative procedure,[56] is to produce nitric oxide directly from the nitrosamine by heating in the gas phase at *ca* 300 °C.

$$\text{Rate} = kh_0[\text{Nitrosamine}][Y^-] \qquad (5.27)$$

Denitrosation has been studied kinetically for a number of nitrosamines, mostly aryl-N-nitrosamines, with a range of nucleophiles Y^- and in several solvents. In water the rate equation (equation (5.27)) has been established, for reaction in the presence of nitrite traps, to ensure the irreversibility of the process.[57] So long as the concentration of the nitrite trap exceeds a certain minimum threshold value (which is different for each trap) then the reaction is truly zero-order in the nitrite trap. Species used include hydrazine, hydrazoic acid, sulphamic acid, hydroxylamine, ascorbic acid, aniline and urea. All of the kinetic results, including the observation of a kinetic isotope effect k_{H_2O}/k_{D_2O} of *ca* 0.3, are consistent with the mechanism set out in scheme (5.28), involving a rapid reversible *N*-protonation followed by a rate-limiting attack by the nucleophile Y^- giving the nitrosyl derivative YNO which is rapidly destroyed with a nitrite trap in these experiments. As

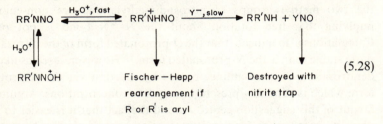

(5.28)

expected the reactivity of the nucleophile parallels that found for other nucleophilic reactions eg S_N2 substitutions. A range of nucleophiles together with their relative reactivities (relative to chloride ion) is given in table 5.5.[57,58] This table includes a range of neutral sulphur nucleophiles RSH, RSCH$_3$ as well as thiourea derivatives. Interestingly there is quite a

Table 5.5. *Relative reactivity of a range of nucleophiles towards N-methyl-N-nitrosoaniline in acid solution*

Nucleophile	Relative reactivity ($Cl^- = 1$)
Alanine	0
Chloride ion	1
Cysteine	2
Glutathione	3
S-Methylcysteine	35
Bromide ion	55
Methionine	65
Thiocyanate ion	5 500
Thiourea	13 000
Methylthiourea	14 250
N,*N*′-Dimethylthiourea	14 500
Trimethylthiourea	12 250
Tetramethylthiourea	13 500
Iodide ion	15 750

good correlation between $\log k$ (equation (5.27)) and the Pearson nu-cleophilicity parameter n^{59}, for those nucleophiles for which n has been measured from S_N2 rate data of methyl iodide. Figure 5.3 shows the relationship for two nitrosamines N-methyl-N-nitrosoaniline and N-nitrosodiphenylamine for the nucleophiles chloride ion, bromide ion, thiocyanate ion, thiourea, and iodide ion. It would seem that the Pearson n values can therefore be taken as quite a good measure of nucleophilicity even for attack at nitrogen.[57,60] In the absence of any added nucleophiles it is believed that in aqueous solution the solvent acts as the nucleophile, thus effecting a hydrolysis of the nitrosamine to give the secondary amine and free nitrous acid. This reaction is necessarily slower than the nucleophile-

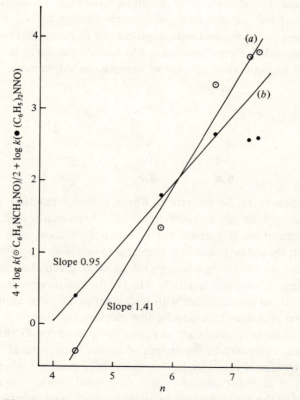

Figure 5.3 The reactivity of nucleophiles in the denitrosation of nitrosamines in terms of the Pearson nucleophilicity parameter n: (a) N-methyl-N-nitrosoaniline and (b) N-nitrosodiphenylamine.

catalysed reactions, but nevertheless can be readily accomplished for aryl nitrosamines, whereas for dialkyl nitrosamines fairly concentrated acids are required.

Direct nitrosation of nucleophiles by nitrosamines is thus clearly established, but is limited to fairly powerful nucleophiles and also to non-basic nucleophiles because of the fairly strong acid concentrations required. There is one report of denitrosation brought about by azide ion in the case of *N*-acetyl-*N'*-nitrosotryptophan,[61] which is so reactive that it can be denitrosated at low acidity, typically pH 6, where substantial quantities of free azide ion exist. Its reactivity is comparable with that of thiocyanate, ie is rather greater than is predicted by the Pearson *n* parameter, which gives similar values to azide ion and bromide ion.

Denitrosation of some aliphatic heterocyclic nitrosamines has also been shown to occur, and some such reactions have been studied kinetically, using an added nitrite trap to ensure irreversibility. 4-Methyl-1-nitrosopiperazine (**5.6**), mononitrosopiperazine (**5.7**), and dinitrosopiperazine (**5.8**), all undergo denitrosation in 3 M hydrochloric acid at 50 °C in the presence of an equivalent amount of ammonium sulphamate.[62]

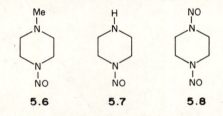

5.6   **5.7**   **5.8**

There appears to be no report where a direct reaction between a nitrosamine and an alcohol has specifically been established, although when denitrosation is carried out in ethanol solution, ethyl nitrite is formed.[63] It should be possible in principle to establish whether the direct reaction can occur, though it appears unlikely given the rather feeble nucleophilicity of alcohols generally. Most of the kinetic work involving the direct reaction of nitrosamines with nucleophiles has been limited to aryl nitrosamines and some heterocyclic nitrosamines, but there is no reason to suppose that these reactions are anything but general for all nitrosamines, although reactivity will be dependent of course on structural factors.

A rather unusual reaction, however, has been detected whereby *direct* nitrosation of aromatic amines occurs. In general, amines are extensively protonated in the acid solutions required to effect denitrosation, and so a direct reaction with the free amine acting as a nucleophile is not to be expected. However, for the reaction of *N*-nitrosodiphenylamine with *N*-

methylaniline in acid solution N-methyl-N-nitrosoaniline is formed, but in a process which is not catalysed by chloride, bromide or thiocyanate ion, although under the experimental conditions the denitrosation of N-nitrosodiphenylamine is markedly catalysed by these anions.[64] The reaction is reversible and kinetic data were limited to initial rate measurements. Later, to avoid the complication of reversibility and to eliminate any indirect reaction, the same nitrosamine was treated with a number of aniline derivatives in acid solution in the presence of an excess of the nitrous acid trap, hydrazoic acid.[60] In each case the corresponding diazonium ions were formed (see equation (5.29)) and the reaction was kinetically first-order in both the aniline derivative and nitrosamine concentrations. Since the reaction is also acid-catalysed (in the range 0.6–3.0 M sulphuric acid) it suggests that the active species is the anilinium ion rather than the free base form of the amine. The reactivity of the anilinium ion itself is comparable with that of chloride ion, with the following sequence of substituent effects $4\text{-}CH_3 > 3\text{-}OCH_3 > H > 4\text{-}Cl$. Aliphatic amines show no such reaction. This reaction is similar to that reported for the diazotisation of anilines in 3 M perchloric acid.[65] The kinetic form is identical and the substituent effects run in parallel, suggesting a similar mechanism. It has been proposed (see Chapter 4) that the nitrosating agents, nitrous acidium ion (or nitrosonium ion) and the protonated form of N-nitrosodiphenylamine attack the anilinium ion giving an intermediate where the nitrosonium ion group is bound to the aromatic ring, after which rearrangement occurs to the amino nitrogen atom.

$$(C_6H_5)_2\overset{+}{N}HNO \; + \; \underset{X}{\overset{\overset{+}{N}H_3}{\bigcirc}} \quad \xrightarrow[HN_3]{H_3O^+} \quad (C_6H_5)_2NH \; + \; \underset{X}{\overset{\overset{+}{N_2}}{\bigcirc}} \tag{5.29}$$

Although these denitrosation reactions have been represented by scheme (5.28) and rate equation (5.27), a variety of experimental conditions exist where a different rate law is observed, given in equation (5.30), where the dependence upon the nucleophile has now disappeared. This occurs generally at high nucleophile concentration, and particularly for the more powerful nucleophiles, for nitrosamide substrates and others containing powerful electron-withdrawing groups, and also for reaction in non-aqueous solvents, typically ethanol. An intermediate situation[64] exists between equations (5.27) and (5.30), and the experimental results are shown graphically in figure 5.4 for the reaction of N-methyl-N-nitrosoaniline in

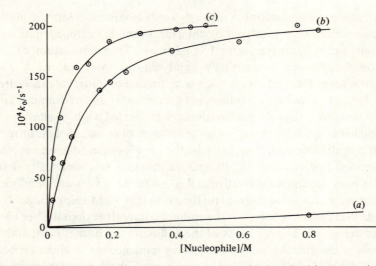

Figure 5.4 Catalysis of denitrosation by (a) bromide ion (b) thiocyanate ion and (c) thiourea.

aqueous acid with the nucleophiles bromide ion, thiocyanate ion, and thiourea.[66] Clearly equation (5.27) applies for bromide ion whereas for

$$\text{Rate} = k[\text{H}_3\text{O}^+](\text{or } h_0)[\text{Nitrosamine}] \qquad (5.30)$$

both thiocyanate ion and thiourea, at low concentration equation (5.27) is followed, and at high concentration we arrive at a zero-order nucleophile dependence. Similarly there are a number of reports of the absence of a kinetic nucleophile dependence in the denitrosation of nitrosamides,[67] a nitrososulphonamide[68] and nitrosoureas.[69,70] Also the rate of denitrosation of N-methyl-N-nitrosoaniline in ethanol solvent is independent of added nucleophiles. An explanation,[69] which covers each of these observations, retains the outline mechanism given in scheme (5.28) but now has the rate-limiting step as the proton transfer to the nitrosamine rather than the attack of the nucleophile Y^-. This will be achieved if, as outlined in scheme (5.31), the inequality $k_2[Y^-] \gg k_{-1}$ applies. Clearly this is more likely at high $[Y^-]$ and for the more powerful nucleophiles, and also in less polar solvents and for nitrosamines containing electron-withdrawing groups such as the carbonyl group. The data given in figure 5.5 have been quantitatively rationalised[64] using the double-reciprocal plot with such a general scheme (scheme (5.31)) for the intermediate situation when $k_2[Y^-] \approx k_{-1}$. Further evidence in support of this suggestion comes from the observation of primary kinetic isotope effects of $k_{\text{H}_2\text{O}}/k_{\text{D}_2\text{O}}$ (or $k_{\text{C}_2\text{H}_5\text{OH}}/k_{\text{C}_2\text{H}_5\text{OD}}$) which lie in the range 1.3–3.8, when the condition

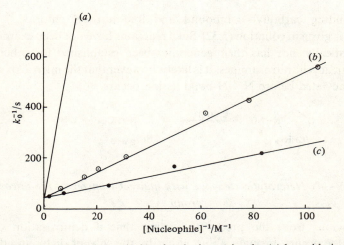

Figure 5.5 Double-reciprocal plots for denitrosation by (*a*) bromide ion, (*b*) thiocyanate ion and (*c*) thiourea.

$k_2[Y^-] \gg k_{-1}$ obtains, contrasting with the value of ~ 0.3 when $k_2[Y^-] \ll k_{-1}$. Also in support of the rate-limiting proton transfer is the observation of general acid catalysis for the denitrosation of nitrosamides[67] and a nitrososulphonamide.[68]

$$RR'NNO + H_3O^+ \underset{k_{-1}}{\overset{k_1}{\rightleftharpoons}} RR'\overset{+}{N}HNO \xrightarrow{Y^-,k_2} RR'NH + YNO \quad (5.31)$$

The change of rate-limiting step achieved in a number of different ways here should also be achievable for the reverse reaction ie the nitrosation of amines, amides etc. This is discussed more fully in Chapter 4 but it is worth noting that for the *N*-nitrosation reaction, loss of nucleophilic catalysis occurs at high $[Br^-]$ in the diazotisation of anilines[71] and at high $[Br^-]$ and $[Cl^-]$ in the nitrosation of diphenylamine,[64] and also for the diazotisation in ethanol solvent;[72] it is also well known that the nitrosation of amides proceeds without nucleophilic catalysis[67,69,70,73] (contrasting markedly with the nitrosation of amines). Finally large primary kinetic isotope effects occur when the limiting condition $k_2[Y^-] \gg k_{-1}$ obtains, eg values of 3.5 and 7.9 have been measured for the nitrosation of methylurea[74] and methylacetamide[75] respectively.

Denitrosation can also be effected by reaction of nitrosamines with a range of acid chlorides such as phosgene, chloroacetyl chloride etc. The reaction has been examined synthetically for a range of heterocyclic nitrosamines such as 1, 3-oxazolidine derivatives, when the product is the

corresponding carbonyl compound and free nitrosyl chloride.[76] An
example is given in equation (5.32). Such reactions have not been examined
mechanistically, nor has their generality been established for other, in
particular, acyclic nitrosamines. It is likely, however that the nitroso oxygen
atom is acylated before N—N bond fission occurs.

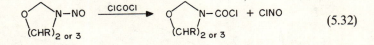

$$(5.32)$$

5.3.2 *N—N Heterolytic cleavage with indirect transfer of the nitroso group*

It is obvious from the preceding section that if denitrosation of a
nitrosamine occurs, brought about either by the solvent or by an added
nucleophile, then the free nitrosating agent, nitrous acid or YNO, generally
is available to nitrosate any other species present in the reaction mixture.
This is the basis of the so-called transnitrosation reactions of nitrosamines,
apart from the few examples noted in section 5.3.1 where a direct transfer of
the nitroso group occurs.

One of the best-known examples occurs in the Liebermann test for a
nitrosamine[77] (or an alkyl nitrite) where it is thought that denitrosation of a
nitrosamine occurs, the free nitrous acid then reacts with phenol to give 4-
nitrosophenol which condenses with a second molecule of phenol to give a
quinone-imine derivative which is coloured bright blue in acid solution and
red in alkaline solution, as outlined in scheme (5.33). This is a transnitrosation to aromatic carbon.

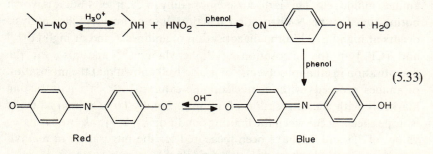

$$(5.33)$$

There are many examples of transnitrosations to amine receptors. Such
reactions are generally reversible and the final product composition
depends on the relative strengths of the N—N bonds in the two
nitrosamines, as well as on the relative concentrations of the reactants. A

large number of aliphatic[78] and alicyclic[79] nitrosamines as well as nitrosoureas[80] have been shown to react in this way, with a range of secondary amines. A few examples are given in equations (5.34), (5.35) and (5.36). In all cases the denitrosation reaction (which can be measured separately in the absence of the added amine and in the presence of a nitrite trap) is rate-limiting. A direct reaction is ruled out by the absence of reaction when no nucleophile such as thiocyanate ion is present. Alternatively, it is possible to demonstrate that transnitrosations of this type occur indirectly by intercepting the free nitrous acid (or its derivative) with some other species which is more reactive, or in higher concentration. This has been shown[61] in the reaction of *N*-acetyl-*N'*-nitroso-tryptophan with 4-nitroaniline (see equation (5.37)) where the 4-nitrobenzenediazonium ion is readily formed. However, when the reaction mixture contains an excess of sulphamic acid, no diazonium ion formation occurs.

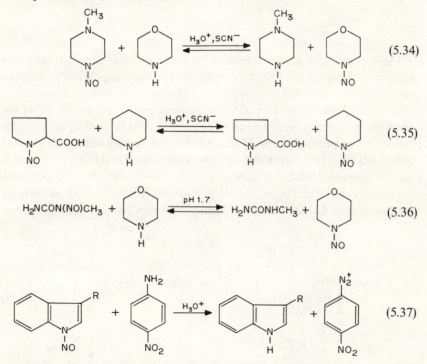

Transnitrosation from nitrosamines to amines has important repercussions in the consideration of the carcinogenic behaviour of nitrosamines. Non-carcinogenic nitrosamines or those with only a weak activity (such as *N*-nitrosodiphenylamine) have the potential to generate more powerfully carcinogenic nitrosamines by transnitrosation *in vivo*, particularly in the acid environment in the stomach, where additional catalysis from naturally occurring nucleophiles might also arise. This topic is discussed in more detail in section 5.4.

As expected, nitrosamines can effect nitrosation of species other than amines, by the indirect mechanism. Thus alcohols are readily converted into alkyl nitrites (see equation (5.38)),[81] but there seems to be no report of thionitrite formation from a nitrosamine and a thiol by this mechanism (although the reverse reaction is known);[82] the issue is obscured by the fact that thiols are sufficiently strong nucleophiles to react directly with nitrosamines, as outlined in section 5.3.1.

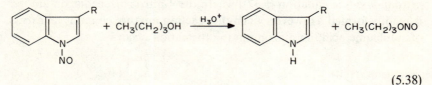

$$(5.38)$$

Indirect nitroso group transfers also occur from nitrosamides and nitrososulphonamides. The deamination pathway competes with denitrosation for nitrosamides, and in all cases the denitrosation pathway is not catalysed by added nucleophiles in contrast to the behaviour of nitrosamines. As for nitrosamines, the free nitrous acid formed upon denitrosation can in principle be used to nitrosate any species. One example (see equation (5.39)) is the quantitative formation of the azo dye derived from 3-hydroxynaphthalene-2, 7-disulphonic acid and the 4-chlorobenzene-diazonium ion, when *N*-methyl-*N*-nitrosourea reacts in acid solution in the presence of 4-chloroaniline.[69] Similarly *N*-methyl-*N*-nitrosotoluene-*p*-

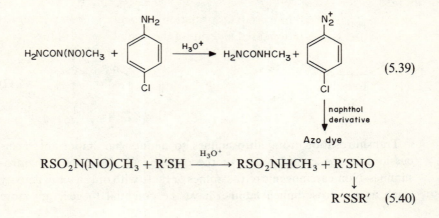

$$RSO_2N(NO)CH_3 + R'SH \xrightarrow{H_3O^+} RSO_2NHCH_3 + R'SNO$$

$$\downarrow$$

$$R'SSR' \quad (5.40)$$

sulphonamide in acid solution reacts with cysteine giving cystine, presumably by intermediate formation and subsequent reaction of *S*-nitrosocysteine (see scheme (5.40)).[83] In this case it is not possible to distinguish between an indirect and direct (quite likely with the nucleophilic thiol) transnitrosation since the rate-limiting step is the initial protonation of the nitrososulphonamide.

5.3.3 N—N Homolytic cleavage with indirect transfer of the nitroso group

The N—N bond in nitrosamines can be broken both thermally and photochemically giving nitric oxide as the primary product. Although itself not a good nitrosating agent towards stable molecules, nitric oxide can, in the presence of oxygen, yield both dinitrogen trioxide and dinitrogen tetroxide which are good nitrosating species. In addition nitric oxide itself can react directly with an organic free radical yielding nitroso compounds.

$$RR'NNO \xrightarrow{\Delta \text{ or } h\nu} RR'N^{\bullet} + NO \tag{5.41}$$

In neutral solution or in the gas phase, thermolysis or photolysis gives the amino radical and nitric oxide (equation (5.41)) but not as readily as in the corresponding reaction of alkyl nitrites (see chapter 6). In fact, nitrosamines are much more stable than are alkyl nitrites, to heat and light (in neutral solution). Consequently much higher temperatures are required to effect homolysis and so the reactions are not so well known, and often lead to a range of products. However, in acid solution reaction occurs more readily, particularly photochemically when it is believed that homolysis of the N—N bond in the *N*-protonated form of the nitrosamine gives the amino radical cation (see equation (5.42)). Neither of these radical species can generally be isolated, but product formation particularly for reaction in acid solution is best rationalised in terms of the formation and subsequent reaction of these intermediates. The acid concentration required is generally quite low, and in fact in > 2 M sulphuric acid reaction does not occur, probably because substantial quantities of the nitrosamines are converted to the unreactive *O*-protonated species under these conditions.[47]

$$RR'NNO \underset{}{\overset{H_3O^+}{\rightleftharpoons}} RR'\overset{+}{N}HNO \xrightarrow{h\nu} RR'NH^{+\bullet} + NO \tag{5.42}$$

Transnitrosation from a nitrosamine to an amine is quite well known. The heterolytic mechanism has already been discussed in section 5.3.2. Reaction can occur readily under quite different experimental conditions, typically by reaction in an organic solvent at *ca* 50 °C or at reflux temperature in the absence of an acid or any other catalyst.[84] The reaction is set out generally in equation (5.43), but has not been fully investigated mechanistically. It has the hallmark of a radical reaction, and esr spectra have been obtained[85] which suggest the involvement of radical intermediates. However, other mechanisms are possible including a direct bimolecular reaction. The indications are[85] that unimolecular homolytic fission of the N—N bond occurs in the rate-limiting step.

$$RR'NNO + R''R'''NH \xrightarrow[\text{organic solvent}]{\Delta} RR'NH + R''R'''NNO \qquad (5.43)$$

A well known example[86] is the reaction of *N*-nitroso-3-nitrocarbazole which converts a number of secondary amines into the corresponding nitrosamines in high yield in benzene solvent under reflux (equation (5.44)).

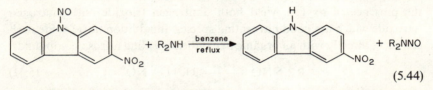

$$(5.44)$$

Transnitrosation to carbon sites has also been used synthetically under conditions which are very similar to those used with amines. Although no mechanistic information is available, again a free-radical mechanism is probable. Thus diphenylnitrosamine converts 1, 2, 3, 4-tetrachloro-cyclopentadiene (equation (5.45)) into the corresponding oxime, by heating in polar solvents such as the lower alcohols or acetic acid.[87] Other substrates containing an 'active methylene group' including pyrazolones, rhodanines, malonic dinitrile, deoxybenzoin and *d*-camphor,[87,88] are similarly converted into the oxime. Other nitrosamines such as dimethylnitrosamine require acid catalysts to effect reaction, presumably to weaken the N—N bond. This can also be achieved by electron-attracting substituents in the 4-positions of the phenyl groups in *N*-nitrosodiphenylamine.

$$(5.45)$$

Nitrosation can also be brought about indirectly in photoaddition reactions to alkenes.[47] An example is given in scheme (5.46) where the amino radical cation attacks the alkene and the resulting free radical forms the *C*-nitroso compound (as the oxime if there are α-hydrogen atoms present) by reaction with the nitric oxide released in the photolysis. Photoadditions of this type are regiospecific, the amino radical cation always attacks the carbon atom resulting in the more stable free radical, which is governed by electronic and steric effects of the substituents. Similar photoaddition reactions of nitrosamines occur with acetylenes[89] and also with conjugated dienes,[90] giving initially the equivalent *C*-nitroso compounds. The photochemistry of nitroso (and nitro) compounds has recently been reviewed by Chow.[91]

$$R_2NNO + H_3O^+ \xrightarrow{h\nu} R_2NH^{+\bullet} + NO$$

(5.46)

$$R_2NH^{+\bullet} + \mathrm{C{=}C} \longrightarrow \mathrm{C{-}C} \xrightarrow{NO} \mathrm{C{-}C}$$

An interesting example of transnitrosation to amines which has some relevance in the rubber industry from the safety viewpoint, concerns the formation of *N*-nitrosodibutylamine and *N*-nitrosodimethylamine from *N*-nitrosodiphenylamine (used as an additive in the rubber industry) and, respectively, dibutylamine and tetramethylthiuran disulphide, both of which are present in rubber mixtures.[92] Both of these nitrosamines are formed under conditions simulating the industrial processes.

5.4. Carcinogenic behaviour of nitrosamines

It seems appropriate in a discussion of nitrosamines as nitrosating reagents to consider also their now generally well known carcinogenic properties, although strictly this topic does not fall within the scope of this chapter. The discovery[93] was made in 1956 that rats, when fed with *N*-nitrosodimethylamine, developed a high incidence of hepatic tumours. Since that time a vast amount of work has been carried out in this area, and it has been shown convincingly that a large number of nitrosamines and nitrosamides are carcinogens, inducing tumours in a wide range of animal species including primates. Indeed all animal species which have been tested are susceptible and the vast majority of *N*-nitroso compounds are reactive in this way, although a small number are non-carcinogens. There is no *direct* evidence which proves that human beings respond in the same way, but it is generally believed that man is not immune. Apart from being carcinogenic, nitrosamines and nitrosamides are generally toxic[94] (effecting liver damage), mutagenic,[95] and teratogenic.[96] Clearly great care is needed when handling these materials either in the laboratory or on a larger industrial scale, to minimise contact, including vapour phase contact. Concern centres around the exposure of man to nitrosamines in the environment generally. Apart from nitrosamine preparations in chemical laboratories and a limited amount of industrial synthesis, it has been established that nitrosamines, generally in low concentrations, are found in a variety of situations in which man may be exposed to them. For example nitrosamines have been detected in some foods, beverages, cosmetics and tobacco products, as well as in the air in certain industrial organisations, notably in the rubber and leather tanning industries. There is a potential health problem here and although no certainty exists as to the adverse effects, particularly of low concentrations of nitrosamines, the situation is constantly being monitored and guidelines issued. The development of the analytical technique of the TEA mentioned earlier means that nitrosamines can now be detected at the level of parts per billion.

Probably a greater exposure to nitrosamines in man arises from endogenous formation rather than from pre-formed nitrosamines. Secondary (and tertiary) amines are widespread in foods, flavourings, tobacco products, a wide variety of drugs etc, and sodium nitrite has been, and is, much used in curing processes for both meats and fish. Further, nitrate ion concentrations in water supplies are increasing, principally owing to the increased use of nitrate-based fertilizers and consequent run-off into rivers. Nitrate is also present in some root and leaf vegetables, and is readily reduced to nitrite by certain bacteria present in saliva and in the stomach.[97] Sodium nitrite is added in meat and fish curing principally as an antibacterial agent against *Clostridium botulinium*, which can produce the very poisonous material botulin, and hence generate botulism. Additionally sodium nitrite has properties which improve the colouring and flavouring in meats. Steps have been taken to reduce the level of nitrite added to meat products, and it is now believed that most of the gastric nitrite derives from nitrate ion and bacterial reduction.

It is to be expected that in the acid environment of the stomach, nitrite and secondary amines would readily generate nitrosamines, particularly in the presence of any catalysts such as chloride ion and thiocyanate ion. It has been shown that nitrosamines are formed *in vitro* with both animal and human gastric juice and also *in vivo* in animals.[98] More recently it has been possible to quantify nitrosamine formation in man by the simultaneous administration of nitrite or nitrate and proline and determining the *N*-nitrosoproline formed (a non-carcinogen) by analysis of the urine.[99] As expected the vitamins C (ascorbic acid) and E (α-tocopherol) are efficient at reducing the amount of nitrosamines formed, by reacting preferentially with nitrous acid.

There are many accounts and discussions of the possible role of nitrosamines in human cancer, for example see appropriate sections in references 2(*a*), (*b*) and 3(*a*), (*b*), (*c*) and also the Banbury report of 1982.[100] Many attempts have also been made to quantify the risk to man from nitrosamines (see references 2(*a*) p. 153, 2(*b*) pp. 207, 217 and reference 101), but there are too many uncertainties at present, for these to lead to any more than very broad and general conclusions. In addition many attempts have been made to correlate incidence of various cancers with nitrite and nitrate levels of population groups (see reference 101 and references therein), but a study[102] has found a surprising *inverse* relation between the nitrite and nitrate levels in the saliva and gastric cancer mortality for some population groups in Britain.

Much effort has been directed at establishing the mechanism of carcinogenesis brought about by nitrosamines. The picture is by no means complete but some features are understood. In common with many other chemical carcinogens it is known that nitrosamines bring about alkylation (believed to be electrophilic) of specific sites within the cellular DNA, which may be responsible for the initiation process in carcinogenesis. The evidence for the *in vivo* reactions of nitrosamines (and other *N*-nitroso compounds) has been reviewed[98] as has the more general area of carcinogenesis brought about by alkylating agents.[103] Briefly, it is thought that nitrosamines first undergo enzymatic α-hydroxylation, and the α-hydroxy nitrosamine breaks down initially to an aldehyde and an unstable primary nitrosamine, from which the alkyl diazonium ion or, in principle, the free alkyl carbocation can readily be formed. This series of steps is outlined in scheme (5.47). Both the alkyl diazonium ion and the alkyl carbocation could react either with water to give alcohol products or with nucleophilic sites in DNA (and also RNA and proteins). This mechanism was first suggested by Druckrey and coworkers[104] and many of its features have been confirmed experimentally, although some of the details are not yet established, for example in the first step of enzymatic hydroxylation; it is likely, however, that a number of enzymes usually present in the liver bring about the transformation. There is quite a lot of support for the intermediacy of the α-hydroxy species; a number of α-acetoxy-dialkylnitrosamines have been synthesised and have been shown to be both carcinogenic[105] and mutagenic.[106] Further these α-acetoxy derivatives are hydrolysed[107] to give product mixtures in accord with the mechanism outlined in scheme (5.47), as shown in equation (5.48). They also effect alkylation in DNA,[108] and are potent direct-acting (that is without any enzymatic transformation) carcinogens.[109] It would appear that the α-hydroxy species itself is extremely unstable, but has been isolated and characterised in one instance.[110]

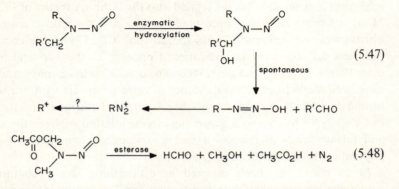

(5.47)

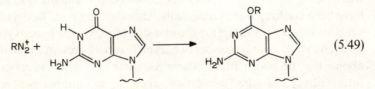

(5.48)

It has been shown that alkylation of DNA under physiological conditions can occur at twelve sites, the ring-nitrogen positions in the bases adenine, guanine, cytosine and thymine and also at the oxygen atoms of hydroxyl or carbonyl groups in guanine, thymine and cytosine, as well as the phosphate group. It is believed that O^6-alkylation in guanine (see equation (5.49)) is one of the critical reactions that occurs,[111] since there is a better correlation of the extent of this O^6-alkylation with carcinogenicity and mutagenicity than with any other alkylations. Such alkylation changes the base-pairing properties, so that during DNA replication, instead of the normal base cytosine, thymine can be incorporated into the newly synthesised DNA strand. A mutational effect of this kind may initiate neoplastic transformation, possibly via activation of cellular oncogenes. A vast amount of research is taking place in this area, which is not appropriately discussed here, the reader is referred to specialised review articles.[111,112,113]

$$RN_2^+ +$$

(5.49)

It appears that the induction of tumours by *N*-nitrosamides does not require enzymatic hydroxylation such as occurs with *N*-nitrosamines, since alkylating species can readily be formed spontaneously at physiological pH values, as outlined in scheme (5.50). The fact that nitrosamides are effective carcinogens at the site of application supports these ideas, whereas the rather pronounced organ specificity of nitrosamines (often at sites remote from that of application), is at least consistent with the view that enzymatic hydroxylation is necessary, which can only occur in those organs capable of effecting such reactions.

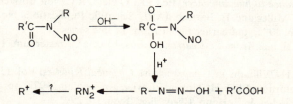

$$(5.50)$$

Before leaving this topic it is worth noting that some nitroso compounds have found application as antitumour agents in chemotherapy. Derivatives of 2-haloethylnitrosoureas (see **5.9**) have been used successfully in this context, after some detailed clinical trials; clearly they have the ability to inhibit tumour growth both in man and in animals. Investigations into their mode of action have concentrated on their spontaneous decomposition reactions under physiological conditions, and subsequent reactions with nucleosides[114] and DNA; the more effective reagents have been found to produce interstrand crosslinks, and this may be the basis of their toxicity.[115]

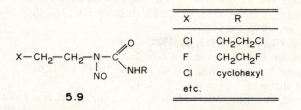

X	R
Cl	CH_2CH_2Cl
F	CH_2CH_2F
Cl	cyclohexyl
etc.	

5.9

References

1. A.L. Fridman, F.M. Mukhametshin and S.S. Novikov, *Russian Chem. Rev.*, 1971, **40**, 34; *The Chemistry of the Nitro and Nitroso Groups*, Part 1, Ed. H. Feuer, Interscience, New York, 1969; B.C. Challis & J.A. Challis in *The Chemistry of Functional Groups Supplement F: The Chemistry of Amino, Nitroso and Nitro Compounds and their Derivatives*, Ed. S. Patai, Interscience, Chichester, 1982, pp. 1151–1208.
2. (*a*) J-P. Anselme (Ed.), *N-Nitrosamines*, ACS Symposium Series 101, ACS, Washington DC, 1979; (*b*) R.A. Scanlan & S.R. Tannenbaum (Eds), *N-Nitroso Compounds*, ACS Symposium Series 174, ACS, Washington DC, 1981.

3. (*a*) *N-Nitroso Compounds: analysis, formation and occurrence*, Eds E.A. Walker, L. Griciute, M. Castegnaro and M. Börzsonyi, IARC Scientific Publications No. 31, IARC, Lyon, 1980; (*b*) *N-Nitroso Compounds: occurrence and biological effects*, Eds H. Bartsch, I.K. O'Neill, M. Castegnaro and M. Okada, IARC Scientific Publications No. 41, IARC, Lyon, 1982; (*c*) *N-Nitroso Compounds: occurrence, biological effects and relevance to human cancer*, Eds I.K. O'Neill, R.C. von Borstel, J.E. Long, C.T. Miller and H. Bartsch, IARC Scientific Publications No. 57, IARC, Lyon, 1985, and earlier publications in this series.

4. H.J. Shine, *Aromatic Rearrangements*, Elsevier, Amsterdam, 1967, pp. 124–271.

5. D.L.H. Williams in *Comprehensive Chemical Kinetics*, Vol. 13, Ed. C.H. Bamford and C.F.H. Tipper, Elsevier, Amsterdam, 1972, pp. 433–66.

6. O. Fischer & E. Hepp, *Chem. Ber.*, 1886, **19**, 2991.

7. O. Fischer & E. Hepp, *Chem. Ber.*, 1887, **20**, 2479; W.J. Hickinbottom, *J. Chem. Soc.*, 1933, 946; J. Willenz, *J. Chem. Soc.*, 1955, 1677.

8. O. Fischer & E. Hepp, *Chem. Ber.*, 1887, **20**, 1247.

9. P.W. Neber & H. Rauscher, *Annalen*, 1942, **550**, 182.

10. W.J. Hickinbottom & S.E.A. Ryder, *J. Chem. Soc.*, 1931, 1281.

11. E. Bamberger, *Chem. Ber.*, 1894, **27**, 1347, 1548.

12. S. Veibel, *Chem. Ber.*, 1930, **63**, 1577.

13. O. Fischer & P. Neber, *Chem. Ber.*, 1912, **45**, 1093.

14. O. Fischer & E. Diepolder, *Annalen*, 1895, **286**, 145.

15. W.L. Bowden, W.F. Little & T.J. Meyer, *J. Am. Chem. Soc.*, 1976, **98**, 444.

16. J. Pasek, A. Jaros, A. Rezabek & M. Mancik, *Czech. CS* 201059; *Chem. Abstr.*, 1984, **99**, 121992.

17. O.K. Nikiforova, *Tr. Khim-Met. Inst. Akad. Nauk SSSR*, 1953, **1**, 53; *Chem. Abstr.*, 1955, **49**, 8158; O. Czeija, *Austrian Pat.*, 163, 203; 1949; *Chem. Abstr.*, 1952, **46**, 9125.

18. J. Houben, *Chem. Ber.*, 1913, **46**, 3984.

19. W. Macmillen & T.H. Reade, *J. Chem. Soc.*, 1929, 585.

20. E.Y. Belyaev & S.V. Petrova, *Zh. Org. Khim.*, 1977, **13**, 1220.

21. D.L.H. Williams, *Tetrahedron*, 1975, 1343.

22. C.K. Ingold, *Structure and Mechanism in Organic Chemistry*, 2nd edn, Bell, London, 1969, p. 901; M.J.S. Dewar, *Molecular Rearrangements*, Ed. P. de Mayo, Interscience, New York, 1963, p. 310.

23. T.I. Aslapovskaya, E.Y. Belyaev, V.P. Kumarev & B.A. Porai-Koshits, *Reakts. spos. org. Soedinenii*, 1968, **5**, 465.

24. D.L.H. Williams & J.A. Wilson, *J. Chem. Soc., Perkin Trans. 2*, 1974, 13.

25. B.T. Baliga, *J. Org. Chem.*, 1970, **35**, 2031.

26. F.G. Soper, *J. Phys. Chem.*, 1927, **31**, 1192; F.G. Soper and D.R. Pryde, *J. Chem. Soc.*, 1927, 2761.

27. T.D.B. Morgan & D.L.H. Williams, *J. Chem. Soc., Perkin Trans. 2*, 1972, 74.

28. D.L.H. Williams & E. Buncel in *Isotopes in Organic Chemistry*, Vol. 5, Eds. E. Buncel and C.C. Lee, Elsevier, Amsterdam, 1980, pp. 147–230.

29. D.L.H. Williams, *J. Chem. Soc., Perkin Trans. 2*, 1975, 655.

30. D.L.H. Williams, *J. Chem. Soc., Perkin Trans. 2*, 1982, 801.
31. T.D.B. Morgan, D.L.H. Williams & J.A. Wilson, *J. Chem. Soc., Perkin Trans. 2*, 1973, 473.
32. G. Ellison & D.L.H. Williams, *J. Chem. Soc., Perkin Trans. 2*, 1981, 699.
33. D.L.H. Williams, *Int. J. Chem. Kinetics*, 1975, **7**, 215.
34. E.Y. Belyaev, V.P. Kumarev & T.I. Aslapovskaya, *Fiz., Khim. Khim. Tekhnol.*, 1969, 321; *Chem. Abstr.*, 1972, **76**, 3201.
35. I.D. Biggs & D.L.H. Williams, *J. Chem. Soc., Perkin Trans. 2*, 1977, 44.
36. I.D. Biggs & D.L.H. Williams, *J. Chem. Soc., Perkin Trans. 2*, 1976, 601.
37. Z. Vrba & Z.J. Allan, *Tetrahedron Lett.*, 1968, 4507.
38. Z.J. Allan, *Tetrahedron Lett.*, 1971, 4225.
39. L.P. Nikitenkova, S.D. Karakotov, Y.A. Strepikheev, B. Kashemirov & I.I. Naumova, *Zh. Obshch. Khim.*, 1981, **51**, 1380 (English translation), and earlier papers.
40. L.P. Kampel, I.I. Naumova, Y.A. Strepikheev & V.I. Zhun, *Zh. Obshch. Khim.* (English translation), 1975, **45**, 2694; L.P. Kampel, Y.A. Strepikheev, I.I. Naumova & L.P. Zharynina, *Zh. Obshch. Khim.* (English translation), 1976, **46**, 2053.
41. J. Thiele, *Annalen*, 1910, **376**, 264.
42. J.R. Perrott, G. Stedman & N. Uysal, *J. Chem. Soc., Perkin Trans. 2*, 1977, 274; G. Stedman & N. Uysal, *J. Chem. Soc., Perkin Trans. 2*, 1977, 667.
43. H. Clusius & K. Schwarzenbach, *Helv. Chim. Acta*, 1959, **42**, 739.
44. R.E. Lyle, W.E. Krueger & V.E. Gunn, *J. Org. Chem.*, 1983, **48**, 3574.
45. H.U. Daeniker, *Helv. Chim. Acta*, 1962, **45**, 2426; 1964, **47**, 33.
46. M. Busch & S. Schäffner, *Chem. Ber.*, 1923, **56**, 1612.
47. Y.L. Chow, *Acc. Chem. Res.*, 1973, **6**, 354.
48. Y.L. Chow, M.P. Lau, R.A. Perry & J.N.S. Tam, *Canad. J. Chem.*, 1972, **50**, 1044.
49. J.N.S. Tam, T. Mojelsky, K. Hanaya & Y.L. Chow, *Tetrahedron*, 1975, **31**, 1123.
50. D. Seebach & D. Enders, *Chem. Ber.*, 1975, **108**, 1293.
51. E.C.S. Jones & J. Kenner, *J. Chem. Soc.*, 1932, 711.
52. S.J. Kuhn & J.S. McIntyre, *Canad. J. Chem.*, 1966, **44**, 105; G.A. Olah, D.J. Donovan & L.K. Keefer, *J. Nat. Cancer Inst.*, 1975, **54**, 465.
53. Y.L. Chow, *Acc. Chem. Res.*, 1973, **6**, 354.
54. T. Sone, Y. Tokuda, T. Sakai, S. Shinkai & O. Monabe, *J. Chem. Soc., Perkin Trans. 2*, 1981, 298.
55. C.L. Walters, R.J. Hart & S. Perse, *Z. Lebensm. Unters. Forsch.*, 1978, **167**, 315.
56. D.H. Fine, D. Lieb & F. Rufeh, *J. Chromatog.*, 1975, **107**, 351.
57. I.D. Biggs & D.L.H. Williams, *J. Chem. Soc., Perkin Trans. 2*, 1975, 107.
58. G. Hallett & D.L.H. Williams, *J. Chem. Soc., Perkin Trans. 2*, 1980, 624.
59. R.G. Pearson, H. Sobel & J. Songstad, *J. Am. Chem. Soc.*, 1968, **90**, 319.
60. J.T. Thompson & D.L.H. Williams, *J. Chem. Soc., Perkin Trans. 2*, 1977, 1932.
61. T.A. Meyer, D.L.H. Williams, R. Bonnett & S.L. Ooi, *J. Chem. Soc., Perkin Trans. 2*, 1982, 1383.

62. S.S. Singer, W. Lijinsky & G.M. Singer, *Tetrahedron Lett.*, 1977, 1613.
63. S.S. Johal, D.L.H. Williams & E. Buncel, *J. Chem. Soc., Perkin Trans. 2*, 1980, 165.
64. B.C. Challis & M.R. Osborne, *J. Chem. Soc., Perkin Trans. 2*, 1973, 1526.
65. E. Kalatzis & J.H. Ridd, *J. Chem. Soc. (B)*, 1966, 529; E.C.R. de Fabrizio, E. Kalatzis & J.H. Ridd, *ibid.*, p. 533.
66. S.S. Al-Kaabi, G. Hallett, T.A. Meyer & D.L.H. Williams, *J. Chem. Soc., Perkin Trans. 2*, 1984, 1803.
67. C.N. Berry & B.C. Challis, *J. Chem. Soc., Perkin Trans. 2*, 1974, 1638; B.C. Challis & S.P. Jones, *ibid.*, 1975, 153.
68. D.L.H. Williams, *J. Chem. Soc., Perkin Trans. 2*, 1976, 1838.
69. G. Hallett & D.L.H. Williams, *J. Chem. Soc., Perkin Trans. 2*, 1980, 1372.
70. J.K. Snyder & L.M. Stock, *J. Org. Chem.*, 1980, **45**, 1990.
71. M.R. Crampton, J.T. Thompson & D.L.H. Williams, *J. Chem. Soc., Perkin Trans. 2*, 1979, 18.
72. A. Woppman & H. Sofer, *Monatsh. Chem.*, 1972, **103**, 163.
73. M. Yamamoto, T. Yamada & A. Tanimura, *J. Food Hyg. Soc. Jpn.*, 1976, **17**, 363.
74. J. Casado, A. Castro, M. Mosquera, M.F.R. Prieto & J.V. Tato, *Ber. Bunsengesellschaft Phys. Chem.*, 1983, **87**, 1211.
75. J. Casado, A. Castro, J.R. Leis, M. Mosquera & M.E. Pena, *Monatsh. Chem.*, 1984, **115**, 1047.
76. K.F. Hebenbrock & K. Eiter, *Annalen*, 1972, **765**, 78.
77. H. Decker & B. Solonina, *Chem. Ber.*, 1902, **35**, 3217; F. Feigl, *Spot Tests in Organic Analysis*, 5th edn, Elsevier, London, 1956, p. 154.
78. S.S. Singer, *J. Org. Chem.*, 1978, **43**, 4612.
79. S.S. Singer, G.M. Singer & B.B. Cole, *J. Org. Chem.*, 1980, **45**, 4931.
80. S.S. Singer & B.B. Cole, *J. Org. Chem.*, 1981, **46**, 3461.
81. R. Bonnett & P. Nicolaidou, *Heterocycles*, 1977, **7**, 637.
82. M.J. Dennis, R.C. Massey & D.J. McWeeny, *J. Sci. Food Agric.*, 1980, **31**, 1195.
83. U. Shulz & D.R. McCalla, *Canad. J. Chem.*, 1969, **47**, 2021.
84. H. Sieper, *Chem. Ber.*, 1967, **100**, 1646.
85. A.J. Buglass, B.C. Challis & M.R. Osborne in *N-Nitroso Compounds in the Environment*, Eds. P. Bogovski and E.A. Walker, IARC Scientific Publications No. 9, IARC, Lyon, 1974, pp. 94–9.
86. C.L. Bumgardner, K.S. McCallum & J.P. Freeman, *J. Am. Chem. Soc.*, 1961, **83**, 4417.
87. C.H. Schmidt, *Angew. Chem. Int. Ed.*, 1963, **2**, 101.
88. D.P. Parihar & S.P. Sharma, *Chem. and Ind.*, 1966, 1227.
89. Y.L. Chow & D.W.L. Chang, *J. Chem. Soc., Chem. Commun.*, 1971, 64.
90. Y.L. Chow, C.J. Colón, H.H. Quon & T. Mojelsky, *Canad. J. Chem.*, 1972, **50**, 1065.
91. Y.L. Chow in *The Chemistry of Functional Groups Supplement F: The Chemistry of Amino, Nitroso and Nitro Compounds and their Derivatives*, Ed. S. Patai, Interscience, Chichester, 1982, pp. 262–81.
92. C. Rappe & T. Rydström in *N-Nitroso Compounds: Analysis, Formation*

and Occurrence, Eds E.A. Walker, L. Griciute, M. Castegnaro and M. Börzsönyi, IARC Scientific Publications No. 31, IARC, Lyon, 1980, 565.

93. P.N. Magee & J.M. Barnes, *Brit. J. Cancer*, 1956, **10**, 114.
94. H.A. Freund, *Ann. Intern. Med.*, 1937, **10**, 1144.
95. R. Montesano & H. Bartsch, *Mutation Res.*, 1976, **32**, 179.
96. L. Tomatis & U. Mohr, Eds., *Transplacental carcinogenesis*, IARC Scientific Publications No. 4, IARC, Lyon, 1973.
97. S.R. Tannenbaum, M. Weismar & D. Fett, *Food Cosmet. Toxicol.*, 1976, **114**, 549.
98. P.N. Magee, R. Montesano & R. Preussmann in *Chemical Carcinogens*, Ed. C.E. Searle, American Chemical Society, Washington DC, 1976, p. 491; J.S. Wishnok, *J. Chem. Educ.*, 1977, **54**, 440.
99. H. Ohshima & H. Bartsch, *Cancer Res.*, 1981, **41**, 3658.
100. P.N. Magee Ed., *Banbury Report No. 12*, Cold Spring Harbor Laboratory, New York, 1982.
101. S.R. Tannenbaum, *Lancet*, 1983, 629.
102. D. Forman, S. Al-Dabbagh & R. Doll, *Nature (London)*, 1985, **313**, 620.
103. P.D. Lawley in *Chemical Carcinogens*, Ed. C.E. Searle, American Chemical Society, Washington DC, 1976, p. 83.
104. H. Druckrey, R. Preussmann, S. Ivankovic & D. Schmahl, *Z. Krebsforsch.*, 1967, **69**, 103.
105. M. Wiessler & D. Schmahl, *Z. Krebsforsch.*, 1976, **85**, 47.
106. M. Okada, E. Suzuki, T. Anjo & M. Mochizuki, *Gann*, 1975, **66**, 457; J.E. Baldwin, A. Scott, S.E. Branz, S.R. Tannenbaum & L. Green, *J. Org. Chem.*, 1978, **43**, 2427.
107. P.P. Roller, D.R. Shimp & L.K. Keefer, *Tetrahedron Lett.*, 1975, 2065.
108. P. Kleihues, G. Doerjer, L.K. Keefer, J.M. Rice, P.P. Roller & R.M. Hodgson, *Cancer Res.*, 1979, **39**, 5136.
109. S.R. Joshi, J.M. Rice, M.L. Wenk, P.P. Roller & L.K. Keefer, *J, Natl Cancer Inst.*, 1977, **58**, 1531.
110. M. Mochizuki, T. Anjo & M. Okada, *Tetrahedron Lett.*, 1980, 3693; see also reference 3(*a*) pp. 71–82.
111. A.E. Pegg, *Adv. Cancer Res.*, 1977, **25**, 195.
112. G.P. Margison & P.J. O'Connor in *Nucleic Acid Modification by N-nitroso compounds*, Ed. P. L. Grover, CRC Press, Boca Raton, 1979, pp. 111–59; A.E. Pegg in *Reviews in Biochemical Toxicology*, Eds E. Hodgson, J.R. Bend and R.M. Philpot, 1983, Vol. 5, pp. 83–133, Elsevier North Holland Biochemical Press, Amsterdam.
113. D. Yarosh, *Mutation Res.*, 1985, **145**, 1.
114. G.A. Digenis & C.H. Issidorides, *Bioorganic Chemistry*, 1979, **8**, 128–32.
115. K.W. Kohn, *Cancer Res.*, 1977, **37**, 1450.

6

O-Nitrosation and alkyl nitrites

6.1 Nitrosation of alcohols

O-Nitrosations are much less commonly encountered than are N- or C-nitrosations and consequently have been much less studied mechanistically. The obvious and by far the best-known example is the formation of an alkyl nitrite from an alcohol and nitrous acid (see equation (6.1)). This reaction is the most commonly used method of preparation of alkyl nitrites.[1] The reaction is quite general for any alkyl group R, but is unknown for phenols, where aromatic substitution by the nitroso group generally occurs. The reaction is reversible, and the position of equilibrium is dependent upon the electronic and steric properties of R. The boiling-points of alkyl nitrites are significantly lower than are those of the corresponding alcohols, so that even though reaction is not quantitative, the alkyl nitrite can be removed by distillation. The reaction is also quite general for a range of nitrosating agents; a common procedure uses a solution of nitrosyl chloride in pyridine[2] but nitrosonium salts[3] and alkyl nitrites themselves[4] have also been used, although a number of these alternatives seem to have no particular experimental advantage over the more conveniently available nitrous acid, generated from sodium nitrite and a mineral acid. It is clear, however, that any electrophilic nitrosating species XNO is capable of effecting nitrosation (see equation(6.2)). The equilibrium constant is also solvent dependent, and the one possible practical advantage of using alkyl nitrites as the source of the nitrosonium ion lies in the ability of such reactions so called transnitrosations, (see equation (6.3)) to take place in non-aqueous solvents, in cases where there might be solubility or other problems in water. Again the experimental procedure has usually involved the removal of R'ONO by distillation or (for R' = CH$_3$ or C$_2$H$_5$) by spontaneous gaseous evolution. The reactions are generally rapid and are very conveniently carried out in the laboratory. A modification[5] using R = t-C$_4$H$_9$ has significant advantages in terms of the position of the equilibrium. This reaction will be discussed more fully later in this chapter (section 6.5) in a more general discussion of the use of alkyl nitrites as nitrosating species.

$$ROH + HNO_2 \rightleftharpoons RONO + H_2O \qquad (6.1)$$

$$ROH + XNO \rightleftharpoons RONO + HX \qquad (6.2)$$

$$RONO + R'OH \rightleftharpoons ROH + R'ONO \qquad (6.3)$$

That equation (6.1) represents *O*-nitrosation rather than nitrite ion attack, has been demonstrated, particularly by the observation by Allen[6] that an optically active alcohol yields the corresponding alkyl nitrite without racemisation. Similarly the reverse reaction, the hydrolysis of the alkyl nitrite occurs by N—O bond fission, again as shown the retention of configuration, and also by the absence of an excess of ^{18}O in the alcohol produced when the hydrolysis is carried out in ^{18}O-enriched water. This means that chiral alkyl nitrites can readily be formed directly from the corresponding chiral alcohol. Procedures in the literature include the use of nitrosyl chloride in pyridine[7] and an alkyl nitrite transnitrosation.[8]

$$K = \frac{[RONO]}{[ROH][HNO_2]} \qquad (6.4)$$

Equilibrium constants for alkyl nitrite formation (defined by equation (6.4)) have been measured for a number of simple alcohols by three different procedures; (*a*) by direct spectrophotometric measurement, (*b*) by an indirect kinetic method, and (*c*) by a direct measurement of the rate constants for the forward and reverse reactions. In method (*a*) measurements were carried out of the absorbance due to the alkyl nitrite in the uv spectrum at 265 nm.[9] Method (*b*) involved the measurement of the rate constants for the nitrosation of morpholine in the presence of varying amounts of the alcohol,[9] when significant amounts of nitrous acid are 'tied

Table 6.1. *Equilibrium constants for the formation of alkyl nitrites*

R	Method (*a*) (25 °C)	$K/l\,mol^{-1}$ Method (*b*) (25 °C)	Method (*c*) (0 °C)
CH_3	3.5 ± 0.1	5.1 ± 0.2	2.5 ± 0.5
C_2H_5	1.20 ± 0.06	1.39 ± 0.04	0.81 ± 0.02
$n\text{-}C_3H_7$	1.3 ± 0.1	1.42 ± 0.04	0.66 ± 0.03
$i\text{-}C_3H_7$	0.52 ± 0.05	0.56 ± 0.03	0.25 ± 0.03
$2\text{-}C_4H_9$	0.46 ± 0.03	0.39 ± 0.02	–
$i\text{-}C_4H_9$	1.53 ± 0.05	1.90 ± 0.02	–
$t\text{-}C_4H_9$	–	–	< 0.05

up' as inactive, alkyl nitrites. In method (c) the rate constants were measured by stopped-flow spectrophotometry with the alcohol concentration in a large excess over that of the nitrous acid, and the problem treated as a reversible reaction first-order in both directions.[10] Some of the results are given in table 6.1 and show generally reasonable agreement between the various methods, given that method (c) refers to 0 °C, whereas the other two procedures were carried out at 25 °C. Clearly the value of K is not very dependent on the structure of the alcohol, values for other alcohols including di and trihydric alcohols, and also for some carbohydrates are also in the same general area.[10] It is however, noteworthy that K decreases along the series $R = CH_3 > C_2H_5 > i - C_3H_7 > t - C_4H_9$ implying that steric effects rather than electronic effects are more important in *O*-nitrosation of alcohols.

In general the nitrosation of alcohols is a rapid process and rate constants have been determined using special techniques for fast reactions. There are no complications involving protonation equilibria of the substrate as is the case for many amines, since over the acidity range studied the extent of *O*-protonation is negligible. For the case of alcohol nitrosation in aqueous solution the plot of the observed first-order rate constant (with $[ROH] \gg [HNO_2]$) against $[ROH]$ is linear, and the slope gives the rate constant for the forward reaction (*O*-nitrosation) and the intercept, that for the reverse reaction (alkyl nitrite hydrolysis). Both reactions are subject to acid catalysis. Some values of the rate constants are given in table 6.2. The trend in the K values for the series methanol, ethanol, *i*-propanol and *t*-butanol given in table 6–1 can now be seen to arise from the decreasing

Table 6.2. *Rate Constants for the nitrosation of alcohols ROH (k_1) and for the hydrolysis of alkyl nitrites RONO (k_{-1}) at $0°C$*

Alcohol	$k_1/l^2\,mol^{-2}\,s^{-1\,a}$	$k_{-1}/l\,mol^{-1}\,s^{-1\,b}$
Methanol	73 ± 10	31 ± 6
Ethanol	38 ± 0.3	47 ± 0.2
n-Propanol	29 ± 1	44 ± 1
i-Propanol	11 ± 1	44 ± 2
t-Butanol	—	*ca* 100
Glycerol	52 ± 10	46 ± 7
Sucrose	100 ± 10	65 ± 2
Glucose	124 ± 18	81 ± 3

[a]Third-order rate constants defined by Rate $= k_1[ROH][HNO_2][H^+]$.
[b]Second-order rate constants defined by Rate $= k_{-1}[RONO][H^+]$.

values of k_1, since the k_{-1} values are fairly constant, which again supports an argument based on steric effects.

The reaction mechanism appears to be straightforward and is set out in scheme (6.5), assuming that the nitrosating species is the nitrous acidium ion, but a similar scheme involving the nitrosonium ion is equally valid on the kinetic evidence. There is no evidence for a discrete intermediate made up of the alcohol and the nitrous acidium ion, but this point has not been probed mechanistically.

The reactivity of the alcohols here is defined, as it is for a number of other substrates in nitrosation by the third-order rate constant defined in table 6.2. Since the pK_a of the nitrous acidium ion is not known, it is not possible to extract the bimolecular rate constant for attack by nitrous acidium ion. For very reactive species, however, values of k_1 show a tendency to level off at a limiting value which is taken to be the encounter rate between the two species.[11] At 25 °C this limit is $ca\ 7000\ l^2\ mol^{-2}\ s^{-1}$ which is realised for a number of aniline derivatives and for thiourea and its derivatives. For methanol the value of k_1 at 25 °C is $700\ l^2\ mol^{-2}\ s^{-1}$, which puts its reactivity at slightly below that of the encounter limit. Other alcohols studied are somewhat less reactive.

Although exact comparisons are not possible, it is quite clear that the reactivity of alcohols and amines in nitrosation are at least comparable even when allowance is made for the protonation of the amines. Thus we have an example here where the basicities and nucleophilicities do not run in parallel. Again no direct comparisons are possible but the basicities of the amines are many orders of magnitude greater than those of alcohols generally and yet the amines and alcohols are of comparable reactivity in the nitrosation reaction. The HSAB principle[12] designates alcohols and amines generally as hard bases (or nucleophiles) so perhaps it is not unreasonable to find that they have comparable reactivities towards a common electrophile.

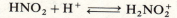

$$HNO_2 + H^+ \rightleftharpoons H_2NO_2^+$$

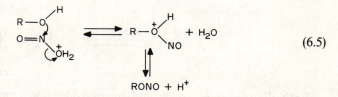

(6.5)

The nitrosation of alcohols using nitrous acid in water is, as expected, catalysed by nucleophilic species, and this has been investigated in detail kinetically for chloride ion and for bromide ion.[10] The degree of catalysis is much less marked than has been found for many amines and other nitrogen-containing species, which is reflected in the values of the bimolecular rate constants k_2 in equation (6.6), which are 2.1×10^5 and $2.0 \times 10^4 \, \text{l} \, \text{mol}^{-1} \, \text{s}^{-1}$ for the nitrosyl chloride and nitrosyl bromide reactions respectively, with methanol in water at 25°C. These contrast with the corresponding values of 2.2×10^9 and $1.7 \times 10^9 \, \text{l} \, \text{mol}^{-1} \, \text{s}^{-1}$ for the reactions of aniline,[13] which are close to the calculated encounter-controlled limit. Again it is possible to rationalise this behaviour in terms of the HSAB principle, since nitrosyl halide is expected to be 'softer' than the positively charged species nitrous acidium ion or nitrosonium ion, binding less favourably with the 'hard' alcohol substrates particularly when compared with the more 'borderline' aromatic amines.

$$\text{ROH} + \text{HalNO} \xrightarrow{k_2} \text{R}-\overset{+}{\text{O}}\!\!\underset{\text{NO}}{\overset{\text{H}}{\diagdown}} + \text{Hal}^- \qquad (6.6)$$

Dinitrogen tetroxide can also be used to produce alkyl nitrites from alcohols. This reagent is capable of both nitrosating and nitrating, depending upon the experimental conditions of temperature and solvent in particular. The role of dinitrogen tetroxide is discussed in more detail in chapter 1. It has been shown that dinitrogen tetroxide can give alkyl nitrites both in the liquid and vapour phase according to equation (6.7).[14] An interesting example of the temperature dependence of the mode of action of dinitrogen tetroxide is shown by its reaction in methylene chloride solution with alkoxide ions;[15] at 0 °C the alkyl nitrite is formed, whereas at -80 °C the alkyl nitrate is the dominant product (see equation (6.8)).

$$\text{N}_2\text{O}_4 + \text{ROH} \longrightarrow \text{RONO} + \text{HNO}_3 \qquad (6.7)$$

$$n\text{-C}_4\text{H}_9\text{O}^-\text{Na}^+ + \text{N}_2\text{O}_4 \quad\begin{array}{l} \xrightarrow{0 \,°\text{C}} \; n\text{-C}_4\text{H}_9\text{ONO} \\ \\ \xrightarrow{-80\,°\text{C}} \; n\text{-C}_4\text{H}_9\text{ONO}_2 \end{array} \qquad (6.8)$$

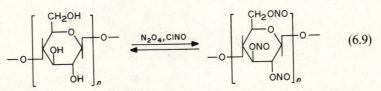 (6.9)

Both dinitrogen tetroxide and nitrosyl chloride react with polysaccharides (eg cellulose) and other carbohydrates to yield equilibrium concentrations of polynitrites. From cellulose itself, the trinitrite is formed (equation (6.9)). The equilibrium position is very solvent dependent, but in basic solvent such as N, N-dialkylamines the trinitrite can be isolated at low temperatures.[16] It is a rather unstable material, decomposing upon warming and also in the presence of mineral acid in a protic solvent, presumably by hydrolysis. Use has been made of this reaction to solubilise cellulose, when clear viscous solutions can be obtained. These polynitrite solutions are readily converted to the more stable sulphates and nitrates, which have been used as thickening agents.[17]

Rate measurements have also been carried out for the nitrosation of alcohols in non-aqueous solvents with nitrosyl chloride. In one study[18] reactions were carried out in acetic acid and a stopped-flow spectrophotometric method used to evaluate the rate constants for a range of alcohols. Here, as is the case for reactions in water, the reactions are reversible but the equilibrium constants are *ca* 10^3 smaller than is the case in water. The kinetics are also more complex than in water, the overall rate equation contains four terms as shown in equation (6.10). The terms containing k_2 and k_{-2} are equivalent to those found in water, whereas the k_3 and k_{-3} terms represent other reaction pathways, which are significant only for the reaction of methanol. Electron-donating substituents increase the value of k_2 by increasing the nucleophilicity of the alcohol. The proposed mechanism involves an adduct formation from the alcohol and nitrosyl chloride. The adduct also includes one molecule of solvent acetic acid or another molecule of reactant alcohol (in the methanol case), bound in a cyclic six-membered ring structure. The reverse reaction, which is first-order in hydrogen chloride, is believed to involve a bimolecular reaction between the alkyl nitrite and hydrogen chloride, and this adduct then reacts with a molecule of either solvent or alcohol (in the methanol case). The situation is rather different from that found with the chloride-ion-catalysed denitrosation reaction in water,[6,10] which is first-order in both $[H^+]$ and $[Cl^-]$, and which will be discussed later in section 6.5.

$$\text{Rate} = k_2[\text{ROH}][\text{ClNO}] + k_3[\text{ROH}]^2[\text{ClNO}]$$
$$- k_{-2}[\text{RONO}][\text{HCl}] - k_{-3}[\text{RONO}][\text{HCl}][\text{ROH}] \qquad (6.10)$$

The reaction of *n*-butanol with nitrosyl chloride in carbon tetrachloride–acetic acid mixtures has also been investigated kinetically by a rather elegant solvent-jump relaxation method.[19] The variation of the rate and equilibrium constants, as well as the activation and thermodynamic parameters, have been rationalised in terms of the bulk properties of the various solvent mixtures.

$$\text{CH}_3\text{OH} + 2\text{NO}_2 = \text{CH}_3\text{ONO} + \text{HNO}_3 \qquad (6.11)$$

Oxides of nitrogen react with alcohols to give a variety of products depending on the conditions. It is worth noting, in the context of this book, that methanol (and other alcohols) reacts with nitrogen dioxide in the gas phase at room temperature to give (equation (6.11)) almost exclusively methyl nitrite, with very little methyl nitrate.[20] Kinetically the reaction is second-order in nitrogen dioxide and first order in methanol, at least at low concentrations, which suggests a nitrosation reaction involving a rate-limiting reaction between dinitrogen tetroxide and methanol.

6.2 Nitrosation of ascorbic acid

Although ascorbic acid is a carbohydrate and therefore an alcohol derivative, its reaction with nitrosating species has been separated from the more general reactions of alcohols and carbohydrates (section 6.1) because of its special biological situation as Vitamin C.

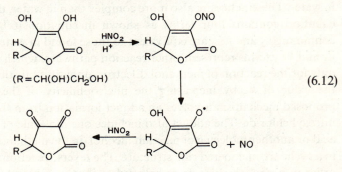

$$(R = CH(OH)CH_2OH) \tag{6.12}$$

Ascorbic acid reacts very readily with two moles of nitrous acid to give dehydroascorbic acid and nitric oxide. A reasonable mechanism (scheme 6.12) involves *O*-nitrosation to give a mononitrite ester, which by homolytic fission gives nitric oxide and a semiquinone radical, which is then rapidly oxidised by another molecule of nitrous acid. Kinetic measurements have shown that at pH < 1 the nitrosating agent is the nitrous acidium ion or the nitrosonium ion, whereas at higher pH values the reactive species is dinitrogen trioxide.[21] At pH 3–5, a significant fraction of the ascorbic acid ($pK_a \sim 4$) exists as the more reactive monoanion, which reacts so rapidly as to make the formation of dinitrogen trioxide rate-limiting. Catalysis by chloride ion, bromide ion and thiocyanate ion is observed as for alcohols generally. Rate measurements on the acid-catalysed reaction show that the reaction of the ascorbate anion is likely to be at the diffusion-controlled limit, with a rate constant very close to that found for reactions of other anions such as azide, acetate and nitrite.[11] As expected, molecular ascorbic acid is less reactive than its anion and has a rate constant which lies between that of hydrazoic acid and some substituted aniline derivatives.[11]

$$HNO_2 + H^+ \xrightarrow{\text{amine}} \text{Nitrosamines etc.}$$

$$\downarrow \text{ascorbic acid}$$ (6.13)

Dehydroascorbic acid

Because of its high reactivity and non-toxic nature, ascorbic acid has been much used to inhibit other nitrosation reactions, eg of amines, in a direct competition (scheme (6.13)). Since the reaction with ascorbic acid is essentially irreversible because of the loss of nitric oxide, the inhibiting effect on other nitrosation pathways is greater than is obtained with alcohols and carbohydrates generally where the *O*-nitrosation is a reversible process. This is of particular relevance to the *in vivo* inhibition of nitrosamine and nitrosamide formation because of their carcinogenic nature. It has been suggested[22] that ascorbic acid be incorporated into drugs containing readily nitrosable secondary amine functions to reduce the extent of *in vivo* nitrosation in the stomach. There is no doubt that added ascorbic acid can suppress or even eliminate amine nitrosation, and there is also much evidence that tumor production in animals resulting from concurrent administration of nitrite and amines or amides, is much reduced or eliminated if ascorbic acid is also fed to the animals.[23] Once nitrosamines have been formed, however, it appears that ascorbic acid cannot play any part in counteracting their carcinogenic behaviour, although it can be used effectively to ensure the irreversibility of the denitrosation of nitrosamines in acid solution.[24] Interestingly, there is a reported inverse correlation between gastric cancer and the consumption of foods containing high concentrations of ascorbic acid,[25] and in regions of the highest incidence of oesophageal cancer, the extent of naturally occurring Vitamin C is very low indeed.[26]

It is clear that ascorbic acid is one of the most suitable of all inhibitors of nitrosation because of its high reactivity, its non-toxicity and also because it is reactive over a wide pH range, because of ionisation to the even more reactive monoanion.

6.3 Nitrosation of carboxylic acids

Just as catalysis of nitrosation by halide ion is interpreted in terms of the intermediacy of the corresponding nitrosyl halides, so the catalysis by added acetate ion is believed to involve the equilibrium formation of nitrosyl acetate (equation (6.14)) and its subsequent reaction with amines etc.[27] This was suggested wholly from the kinetic evidence before nitrosyl acetate had been identified. Subsequently[28] nitrosyl acetate (or acetyl nitrite) has been prepared from solid silver acetate and nitrosyl chloride at liquid nitrogen temperature (equation (6.15)). Nitrosyl acetate is a pale brown liquid at room temperature, a green liquid at $-78\,°C$ and a green solid at $-196\,°C$. Rapid hydrolysis occurs in water, but in a solvent such as pyridine or acetic acid, nitrosyl acetate can be used as an effective nitrosating agent; it gives octyl nitrite from 1-octanol (see equation (6.16)) and a range of products in the deamination of octylamine (see equation (6.17)), very similar to the product distribution obtained when the deamination is carried out using sodium nitrite in acetic acid.

$$HNO_2 + H^+ + CH_3COO^- = CH_3COONO + H_2O \qquad (6.14)$$

$$CH_3COOAg + ClNO \longrightarrow CH_3COONO + AgCl \qquad (6.15)$$

$$CH_3(CH_2)_6CH_2OH + CH_3COONO \xrightarrow{\text{pyridine}} CH_3(CH_2)_6CH_2ONO$$
$$(6.16)$$

$CH_3(CH_2)_6CH_2NH_2$

$+CH_3COONO \longrightarrow$

Octenes	15%
$CH_3(CH_2)_5CH(CH_3)OH$	6%
$CH_3(CH_2)_5CH(CH_3)OOCCH_3$	20%
$CH_3(CH_2)_6CH_2OH$	3%
$CH_3(CH_2)_6CH_2OOCCH_3$	48%

(6.17)

Earlier a range of similar nitrosyl carboxylates had been prepared from perfluorocarboxylic acids[29] using the silver salts and nitrosyl chloride, and also from the reaction of the corresponding acid anhydride with dinitrogen trioxide.[30] These perfluoro derivatives are also unstable, reacting violently with water to give the carboxylic acid and nitrous acid, and decomposing on heating to give the perfluoronitroso alkanes.

It is generally believed that nitrosyl acetate is the reagent involved in nitrosation reactions using sodium nitrite dissolved in acetic acid. Great care is necessary in the interpretation of kinetic results obtained from nitrosation reactions carried out in the presence of carboxylate buffers, because of the complexity of the systems. In a series of papers which laid the foundation for a complete understanding of the mechanism of nitrosation reactions, Hughes, Ingold and Ridd[27] rationalised their observations of catalysis of aniline diazotisation by acetate (and phthalate) buffers in terms of nitrosyl acetate formation. Under the conditions of the experiments the reactions were zero-order in aniline and second-order in nitrous acid, consistent with rate-limiting dinitrogen trioxide formation. Acetate catalysis occurs, as shown in scheme (6.18), by catalysis of dinitrogen trioxide formation rather than by reaction of nitrosyl acetate itself with the amine. Later Stedman[31] showed that in the reaction between nitrous acid and hydrazoic acid in acetate buffers, two parallel reactions occur; in one, nitrosyl acetate is formed from the nitrous acidium ion and acetate ion, which reacts with nitrite ion to give dinitrogen trioxide as for the aniline case (scheme (6.18)), and in the other the nitrosyl acetate so formed reacts directly with azide ion (scheme (6.19)). In other words under these conditions both dinitrogen trioxide and nitrosyl acetate nitrosate azide ion. More recently a Spanish group[32] have examined kinetically the nitrosation of N-methylaniline and piperazine in the presence of an acetate buffer. A detailed analysis shows that, for N-methylaniline, nitrosyl acetate is the only reagent, whilst for piperazine, dinitrogen trioxide and the nitrous acidium ion are also involved. The results suggest that nitrosation by nitrosyl acetate is a diffusion-controlled process, since nitrite ion, N-methylaniline, piperazine and azide ion all react at about the same rate. This enables the equilibrium constant for the formation of nitrosyl acetate (equation (6.20)) to be estimated as $\sim 1.4 \times 10^{-8} \, \text{l mol}^{-1}$. This very small value means that it is virtually impossible to detect nitrosyl acetate in aqueous solution.

$$\left. \begin{aligned} CH_3COO^- + H_2NO_2^+ &\rightleftharpoons CH_3COONO + H_2O \\ NO_2^- + CH_3COONO &\longrightarrow N_2O_3 + CH_3COO^- \end{aligned} \right\} \quad (6.18)$$

$$CH_3COONO + N_3^- \longrightarrow CH_3COO^- + N_3NO \xrightarrow{\text{fast}} N_2 + N_2O \quad (6.19)$$

$$CH_3COOH + HNO_2 = CH_3COONO + H_2O \quad (6.20)$$

$$\text{Rate} = k[HNO_2][CH_3CO_2^-][H^+] \quad (6.21)$$

The rate constant for the formation of nitrosyl acetate (equation (6.21)) was found by an extrapolative procedure, from the variation of the rate constant with azide ion concentration,[31] to be $2200 \, \text{l}^2 \, \text{mol}^{-2} \, \text{s}^{-1}$ at $0 \,^\circ C$. This is very similar to the values found for the rate constants for reactions of a number of other singly charged ions at the same temperature, eg ascorbate ion ($2000 \, \text{l}^2 \, \text{mol}^{-2} \, \text{s}^{-1}$), azide ion ($2340 \, \text{l}^2 \, \text{mol}^{-2} \, \text{s}^{-1}$), nitrite ion ($1893 \, \text{l}^2 \, \text{mol}^{-2} \, \text{s}^{-1}$), etc., and this is taken to represent the diffusion-controlled limit for the reaction of each anion with the nitrous acidium ion or the free nitrosonium ion.

6.4 Nitrosation of hydrogen peroxide

Hydrogen peroxide reacts very readily with nitrous acid in acid solution to give nitric acid via, it is believed, peroxynitrous acid HOONO (see scheme (6.22)). Both reactions are quite rapid with the formation of peroxynitrous acid being the faster. The involvment of this intermediate was first suggested by Gleu and Hubold,[33] and later its absorption spectrum was measured.[34] By stabilising the species as its anion in alkaline solution, it proved possible to isolate the sodium salt of peroxynitrous acid but not in a highly pure form because of its inherent instability.[35] Apart from its spontaneous decomposition to nitrate ion, peroxynitrous acid can be used both to hydroxylate and to nitrate aromatic compounds,[36] but otherwise, not much is known about its chemistry. The mechanism of its formation, however, appears to be a quite straightforward example of *O*-nitrosation similar to that of alcohols. The rate equation (equation (6.23)) has been established experimentally,[37] and is identical to that found for a range of substrates where the nitrous acidium ion or the nitrosonium ion are the active species. Tracer experiments with ^{18}O have shown[38] that both oxygen atoms from the hydrogen peroxide molecule end up in the nitrate anion. The outline mechanism given in scheme (6.24) accounts for the experimental facts. A more recent kinetic study[39] has established the rate equations both for the formation and for the subsequent decomposition of peroxynitrous acid by stopped-flow spectrophotometry. This study confirmed the earlier findings, but found that values of $k_1/[H^+]$ (where k_1 is the observed first-order rate constant for the disappearance of nitrous acid when $[H_2O_2] \gg [HNO_2]$) tended to level off at high concentrations of hydrogen peroxide. This was taken to indicate a change of rate-limiting step to the rate-limiting formation of the nitrosonium ion. Similar behaviour has been found for the nitrosation of alcohols[40] (where the limiting value very much depends on the nature of the alcohol) which was interpreted as a medium effect rather than a change to a true zero-order dependence upon the substrate concentration. Rather high concentrations of hydrogen peroxide (*ca* 1 M) are required to bring about the change, similarly again to the alcohols reactions, and it may well be that the kinetic observations here also result from a medium effect.

$$\left. \begin{array}{l} HNO_2 + H_2O_2 \xrightarrow{\;H^+\;} HOONO + H_2O \\ HOONO \longrightarrow NO_3^- + H^+ \end{array} \right\} \qquad (6.22)$$

$$\text{Rate} = k[HNO_2][H_2O_2][H^+] \qquad (6.23)$$

Irrespective of the detailed mechanism of nitrosation of hydrogen peroxide, it is clear that hydrogen peroxide is one of the most reactive of substrates. The third-order rate constant (equation (6.23)) has a value of *ca* $840 \, l^2 \, mol^{-2} \, s^{-1}$ at 0 °C (taken from the initial slope of $k_1/[H^+]$ *vs* $[H_2O_2]$), which is quite close to the value of $640 \, l^2 \, mol^{-2} \, s^{-1}$ found for thiourea,[41] but is rather less than the value of $3000 \, l^2 \, mol^{-2} \, s^{-1}$ quoted for azulene at the same temperature.[42] It seems very likely that reaction occurs at or very close to the diffusion-controlled limit. There is no report of the use of hydrogen peroxide as an inhibitor of eg amine nitrosation, or its use as a nitrite scavenger or trap. It would appear to be particularly suited for that role.

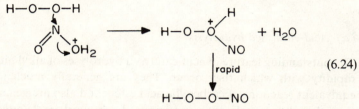

$$(6.24)$$

In the pH ranges 12–13 and 4–9[35,43] the decomposition of peroxynitrous acid is first-order in [HOONO] but is independent of the pH of the solution. A kinetic analysis of the results from the 4–9 pH range gave $pK_a = 6.6$ for peroxynitrous acid.[43] A more extensive investigation by Benton and Moore[39] showed that there is also an acid-catalysed pathway for the decomposition which would not be observable when pH > 2. This suggests that the decomposition occurs by two parallel pathways (see scheme (6.25)) one of which involves the neutral molecule and the other its protonated form. No further mechanistic details have been established. Similarly tertiary hydroperoxides give nitrates[44] in reaction with nitrosyl chloride, probably also by *O*-nitrosation to give initially the peroxynitrite, which spontaneously rearranges to the nitrate (see equation (6.26)). Dinitrogen tetroxide reacts in the same way with hydroperoxides in organic solvents.[45]

$$
\begin{array}{l}
\text{HOONO} \xrightarrow{\hspace{1cm}} NO_3^- + H^+ \\[4pt]
H^+ \Big\updownarrow \hspace{5.5cm} (6.25) \\[4pt]
H_2\overset{+}{O}ONO \xrightarrow{\hspace{1cm}} NO_3^- + 2H^+
\end{array}
$$

$$(CH_3)_3COOH \xrightarrow{\text{ClNO}} (CH_3)_3COONO \longrightarrow (CH_3)_3CONO_2 \quad (6.26)$$

6.5 Reactions of alkyl nitrites

This section will deal only with reactions of nitrite esters where they act as nitrosating agents, ie where the nitroso group is transferred either intermolecularly or intramolecularly. Alkyl nitrites do, of course, undergo a number of other important reactions, involving both homolytic and heterolytic processes, but these are not within the scope of this book.

6.5.1 *Hydrolysis of alkyl nitrites*

Alkyl nitrites undergo both acid-catalysed and base-catalysed hydrolysis. These can be thought of as nitrosation of water, but are also important reactions in the context of the other nitrosation reactions of alkyl nitrites.

(a) Acid-catalysed hydrolysis

The outstanding feature of acid-catalysed hydrolyses of alkyl nitrites is the rapidity with which they occur. They are generally much faster than equivalent reactions of carboxylic acid esters and also are generally faster than their base-catalysed counterparts. Early workers[46] found that the reactions were too fast to measure by conventional techniques. Later it has been found possible to measure rate constants for the hydrolysis of a number of structurally different nitrites. Allen[6] established the rate law given in equation (6.27) for the hydrolysis of both propyl nitrite and *t*-butyl nitrite in 72% dioxan-water at 0 °C. As noted earlier in section 6.1 the reactions are generally reversible. Rate constants for the hydrolysis of alkyl nitrites were also obtained simultaneously with those for the nitrosation of alcohols in water, by treating the system as a reversible kinetic experiment. The rate constants for a number of alkyl nitrites are given in table 6.2 and show very little change with structure. This suggests that reactions occur between the protonated form of the alkyl nitrite and a water molecule (equation (6.28)), at or close to the encounter limit. This in turn would suggest $pK_a \approx -8$ for eg methyl nitrite, which is not an unreasonable value. Any reactions of alkyl nitrites in water are thus accompanied by hydrolysis, and it has to be established in each case whether subsequent reaction occurs either by way of the intermediate nitrous acid formed, or by a direct reaction involving the alkyl nitrite molecule. As indicated earlier in section 6.1, the reactions are reversible and equilibrium constants have been determined. A noteworthy feature is the situation for tertiary nitrites

$$\text{Rate} = k[\text{H}^+][\text{RONO}] \qquad (6.27)$$

$$\qquad (6.28)$$

such as *t*-butyl nitrite, where hydrolysis proceeds virtually to completion. This has been interpreted in terms of a steric effect on the reverse reaction.[10]

The site of protonation of an alkyl nitrite has been assumed to be the oxygen atom bound to the alkyl group. This makes sense in terms of acceptable mechanisms for both the hydrolysis of alkyl nitrites and for the reverse reaction, and is also the site predicted for protonation by molecular orbital calculations.[47,48]

(b) Base-catalysed hydrolysis

This reaction has been much more studied than has the acid-catalysed counterpart. The reaction is essentially irreversible since the nitrous acid produced is present as the inactive nitrite anion. In general reactions are significantly slower than are the corresponding hydrolyses of carboxylic acid ester. As for the acid-catalysed reactions, O—N bond fission occurs as shown by stereochemical experiments,[6] and also by the absence of any excess ^{18}O in the product alcohol when the hydrolysis is performed in $H_2{}^{18}O$ enriched water.[49] The absence of any oxygen exchange between the ester and the hydroxide ion during hydrolysis[49] (in spite of an earlier report to the contrary[50]) suggests that the reaction is a concerted process (equation (6.29)), and does not proceed via an addition–elimination pathway, which is the well-known $B_{Ac}2$ mechanism for the hydrolysis of carboxylic acid esters. The differences in reactivity pattern between alkyl nitrites and carboxylic acid esters, including the different polar and steric effects of substituents have been rationalised[49] in terms of the difference in electronegativity between the carbon and nitrogen, and also in terms of the effect of the lone pair on nitrogen not present in the carboxylic esters.

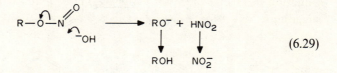

$$(6.29)$$

6.5.2 Nitrosation by alkyl nitrites

Alkyl nitrites can be used as electrophilic nitrosating agents quite generally, reacting with the same range of substrates as do other nitrosating agents. Thus diazonium ions can be generated from primary aromatic amines, nitrosamines from secondary amines, alkyl nitrites from alcohols (transnitrosation), thionitrites from thiols, and a range of *C*-nitrosation reactions can be effected. The advantage of using alkyl nitrites is apparent when substrates of low solubility in water are used, since alkyl nitrites can apparently be used in a range of solvents. Knoevenagel in 1890[51] showed that diazonium ion solutions could readily be produced from alkyl nitrites in a range of solvents such as alcohols, acetic acid, dioxan, tetrahydrofuran etc. In some cases it is possible to precipitate the diazonium salt – a common procedure for isolating these unstable materials, which is generally not possible from aqueous solution using nitrous acid as the nitrosating agent.

The chemistry of diazonium ions is not a subject for this book and is fully covered elsewhere[52] but it is perhaps worth noting a few examples where diazonium ions produced from alkyl nitrites have undergone rather unusual reactions. One of the classic experiments for demonstrating the existence of benzyne involves the diazotisation of anthranilic acid with an alkyl nitrite in aprotic media such as dichloromethane, acetone etc., to give the stable diazonium carboxylate internal salt (equation (6.30)). When heated this decomposes, possibly via a cyclic intermediate (equation (6.31)), to give the very reactive benzyne which is trapped out in the usual way with furan, anthracene etc.[53] A number of interesting reactions with synthetic potential derive from initial diazotisations using alkyl nitrites. For example pentyl nitrite reacts with aniline derivatives in benzene solvent at *ca* 60 °C to yield the aryl phenyl hydrocarbon[54] (see scheme (6.32)), whilst pentyl nitrite with an arylamine in bromoform gives the aryl bromide.[55] Reaction is believed to involve, in both cases, the aryl radical derived from the diazonium ion, since the ratio of isomers obtained is the same as that obtained from standard radical phenylations.

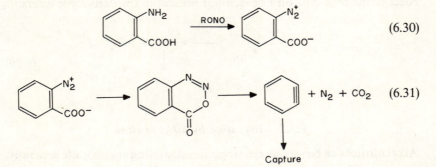

$$RONO + ArNH_2 \longrightarrow ArN_2^+ \longrightarrow Ar^{\bullet} \Big\}$$
$$Ar^{\bullet} + C_6H_6 \longrightarrow ArC_6H_5 \Big\}$$
(6.32)

In aqueous solution or in mixed aqueous organic solvents, reactions of alkyl nitrites particularly in acid solution, are undoubtedly complicated by the concurrent rapid and generally reversible hydrolysis, and it is not easy to establish whether in subsequent nitrosations of eg amines, the effective nitrosating agent is the nitrous acid formed by hydrolysis, or the alkyl nitrite itself. A comparative study of the kinetics of diazotisation of sulphanilamide in aqueous acid solution by both cyclohexyl nitrite and nitrous acid, revealed that both reactions showed a similar kinetic behaviour indicating that reactions do involve the same nitrosating agent probably dinitrogen trioxide.[56]

A more recent kinetic study[57] with 2-propyl nitrite in aqueous acid solution has established this more firmly for the reactions with hydrazoic acid, sulphamic acid, thioglycolic acid and N-methylaniline (a typical range of nitrosatable substrates). The results are quantitatively in agreement with a reversible hydrolysis of the alkyl nitrite, giving free nitrous acid, which effects nitrosation (via nitrous acidium ion or nitrosonium ion) in the rate-limiting stage (see scheme (6.33)). The kinetic analysis, particularly of the reduction of the observed rate constant with added aliphatic alcohol, enabled values of both K (the equilibrium constant for alkyl nitrite hydrolysis) and k_2 (the rate constant for nitrosation with nitrous acidium ion/nitrosonium ion) to be obtained for each substrate S; these values agree well with those measured directly. Butyl nitrite behaves similarly.

$$\left.\begin{array}{c} RONO + H_2O \xrightarrow{\text{H}^+,K} ROH + HNO_2 \\ HNO_2 + S \xrightarrow{\text{H}^+,k_2} Product \end{array}\right\} \tag{6.33}$$

The same pattern occurs for reactions of alkyl nitrites in acid solution in alcohol solvents, at least for the nitrosation of amines. Using propyl nitrite in propanol, nitrosation of aniline and N-methylaniline is very slow in *ca* 0.1 M acid.[58] However, in the presence of chloride ion, bromide ion or thiourea, nitrosation (or diazotisation) takes place very readily. Reaction is first order in [Nucleophile] at low concentration (but is complicated by reversibility of N-nitrosation at higher concentrations of nucleophile), and the experimental results are well accommodated by the mechanism outlined in scheme (6.34), where denitrosation of the alkyl nitrite occurs, by nucleophilic attack of bromide ion etc., at the nitrogen atom in the protonated alkyl nitrite. The nitrosyl bromide (or other nitrosyl compound) then reacts with the amine in the rate-limiting step. For both aniline and N-methylaniline there is no overall acid catalysis, since protonation of the alkyl nitrite is counterbalanced by the amine protonation. But for the less basic 4-nitroaniline, a degree of acid catalysis is found as expected.

$$RONO + H^+ \rightleftharpoons \overset{+}{R}\underset{H}{O}NO \xrightarrow{Br^-} ROH + BrNO \qquad (6.34)$$

slow | amine ↓

Nitrosation products

Direct nitrosation of bromide ion and chloride ion was first reported by Allen,[6] using alkyl nitrites in acidified dioxan-water. The kinetic form of the reactions is given by equation (6.35), the first term represents reaction initiated by the solvent, and the second that brought about by halide ion reaction with the protonated alkyl nitrite. Similar results were obtained from measurements on the equilibrium reaction but starting from the alcohols,[10] which is discussed earlier in section 6.1. In the presence of halide ion, both the forward and reverse reaction rates are increased, bromide ion being only slightly more effective than chloride ion, implying that the bimolecular step ($R\overset{+}{O}(H)NO$ + halide ion) occurs close to the diffusion-controlled limit. This leads to an estimated value of $pK_a \approx -7$ for methyl nitrite, which agrees well with the estimate of ≈ -8 obtained from the solvent promoted reaction. A larger range of nucleophiles has been studied[59] (in dioxan-water at pH 5.5) and the order of reactivity is similar to that found for the nitrous acidium ion/nitrosonium ion reactions ie, azide ion > acetate ion > ascorbate ion > chloroacetate ion > nitrite ion > thiocyanate ion > thiourea > 4-methylaniline > bromide ion > chloride ion > 4-bromoaniline.

The nitroso group exchange between alkyl nitrites and alcohols in acid solution is an example of *O*-nitrosation. Using the alcohol as solvent Allen and Schonbaum[60] found that reaction was reversible, and first-order in both [RONO] and [H$^+$]. The reactive species is again believed to be the protonated alkyl nitrite, which reacts directly (see scheme (6.36)) with the

$$\text{Rate} = k[H^+][RONO] + k'[H^+][RONO][\text{Halide}] \qquad (6.35)$$

$$RONO + H^+ \overset{fast}{\rightleftharpoons} \overset{+}{R}\underset{H}{O}NO$$

$$\overset{+}{R}\underset{H}{O}NO + R'OH \overset{slow}{\rightleftharpoons} [\overset{\delta+}{R}\underset{H}{O} \cdots \underset{O}{\overset{||}{N}} \cdots \overset{\delta+}{\underset{H}{O}R'}]^{\ddagger}$$

$$ROH + R'\overset{+}{\underset{H}{O}}NO \rightleftharpoons R'ONO + H^+ \qquad (6.36)$$

alcohol present in vast excess. There is no evidence for (or against) the involvement of a discrete intermediate, and so the structure in scheme (6.36) is represented as the transition state. Chloride ion catalysis occurs, which suggests that an additional pathway involving nitrosyl chloride then comes into play.

Nitrosation by alkyl nitrites can also occur under alkaline conditions. Amines, alcohols, ketones, nitro compounds, some hydrocarbons, etc., can be nitrosated this way. Thiols also yield thionitrites and dialkyl sulphides give yellow or orange coloured solutions (typical of *S*-nitrosation), although no products have been identified. Synthetically it has been claimed[61] that alkyl nitrites bearing β-electron-withdrawing groups are particularly efficient at nitrosating secondary amines in aqueous alkali (see equation (6.37)). Alkyl nitrites derived from carbohydrates such as glucose or mannose react similarly. These nitrites can either be prepared independently, or generated *in situ* from the corresponding alcohol and nitrosyl chloride (or in principle any nitrosating agent).

$$XCH_2ONO + \underset{}{\rangle}NH \xrightarrow{\ OH^-\ } XCH_2OH + \underset{}{\rangle}NNO \tag{6.37}$$

$$(X = CH_2OC_2H_5, CH_2OH, CH_2F, CF_3)$$

The most widely studied nitrosation reactions under alkaline conditions involve the reaction of amines. Reaction is envisaged as a direct nucleophilic attack by the amine at the nitrogen atom of the alkyl nitrite (see equation (6.38)) to form the nitrosamine, which is stable if the amine is secondary, but which decomposes to give hydrocarbon products (as in the deamination reactions initiated by nitrous acid under acid conditions) if the amine is primary.[62] Tertiary amines can also react. 2-Phenylethyl nitrite with trimethylamine gives the trimethylammonium ion and nitrite ion[63] (equation (6.39)), but it is also possible to bring about nitrosative dealkylation with alkyl nitrites,[64] just as with nitrous acid and tertiary amines (see section 4.3).

$$R-O-N\overset{\displaystyle\nearrow O}{\underset{\displaystyle C:NHR'R''}{}} \longrightarrow ROH + R'R''NNO \tag{6.38}$$

$$C_6H_5CH_2CH_2ONO + (CH_3)_3N \xrightarrow{\ H_2O\ } C_6H_5CH_2CH_2OH$$
$$+ (CH_3)_3\overset{+}{N}H + NO_2^- \tag{6.39}$$

Kinetic measurements of the nitrosation of a number of amines (primary and secondary) in dioxan-water,[63] have revealed that the reactivity of amines is unusually high – typically greater than that of hydroxide ion in the same reaction. For example 2-phenylethyl nitrite reacts 19 times faster with piperidine than it does with hydroxide ion. This has been rationalised in terms of a concerted reaction rather than the more familiar addition–elimination mechanism which occurs for the carboxylic acid ester hydrolyses. Reactivities of the amines do not correlate with their basicities, but rather with their vertical ionisation potentials, indicating that the reactions are orbital-controlled. The important factor is the high electronegativity of nitrogen compared with carbon, so that the lowest unoccupied molecular orbital for an alkyl nitrite is substantially lower in energy than is the corresponding one in a carboxylic acid ester. The solvent isotope effect k_{H_2O}/k_{D_2O} is *ca* 2, implying that proton transfer occurs in the same rate-limiting step, so that the proposed transition state is a cyclic structure, as outlined in scheme (6.40). The variation of the observed rate constant with pH,[65] shows clearly that the reactive species is the free amine and not its protonated form. Ionic strength effects and the value of the entropy of activation support the mechanism given in scheme (6.40).

$$R-O-N\overset{\displaystyle O}{\diagdown} \;+\; H-N\diagup \quad\longrightarrow\quad \left[\begin{matrix} R-\overset{\delta-}{O}-----N\overset{\diagup O}{} \\ | \qquad\quad | \\ H------N\overset{\delta\pm}{\diagdown} \end{matrix}\right]^{\ddagger} \quad\longrightarrow\quad ROH \;+\; \underset{\diagup}{\overset{\diagdown}{}}NNO \qquad (6.40)$$

Nitrosation at oxygen can also occur in alkaline solution. Nitroso group exchange between an alkyl nitrite and an alcohol under these conditions involves nucleophilic attack by the alkoxide ion (equation (6.41)).[66] There is no evidence for the existence of an intermediate species.

$$RONO + R'O^- \rightleftharpoons [R\overset{\delta-}{O}-----\overset{\overset{\displaystyle O}{\|}}{N}-----\overset{\delta-}{O}R']^{\ddagger} \rightleftharpoons RO^- + R'ONO \qquad (6.41)$$

A much less studied area is the nitrosation by alkyl nitrites in non-hydroxylic solvents. In some cases reaction appears to be rapid and quantitative. For example *t*-butyl nitrite is a particularly effective reagent in acetonitrile or chloroform, and provides an excellent synthetic pathway for *N*-, *O*-, and *S*-nitrosation generally.[67] At least in some cases traces of acid are necessary to catalyse the reactions; even small amounts produced by slight decomposition of the alkyl nitrite are sufficient.[66] This suggests that the protonated alkyl nitrite is the reactive species, and is a remarkably reactive nitrosating agent in these solvents. Further weight is added to this suggestion by the observation that the nitroso group exchange between *t*-butyl nitrite and methanol in dichloromethane occurs *ca* 10^3 times more slowly, if the solvent is first passed through an alumina column to remove traces of acid. In general however, kinetic measurements in these solvents have not been reported, and the area needs further study.

Equilibrium constants on the other hand, have been reported for the nitroso group exchange reaction with alcohols, in a range of solvents.[5] Values of K (equation (6.42)) decrease as the structure of the alcohol is changed primary > secondary > tertiary, which can be rationalised in terms of the decreasing nucleophilicities of the alcohols. Consequently tertiary alkyl nitrites are the most effective nitrosating agents, both from the equilibrium viewpoint, and also on kinetic grounds, although there are only few kinetic data in the literature. Both polar and steric effects are also important. Electron-donating groups in the alcohol favour alkyl nitrite formation as expected.

$$K = \frac{[ROH][R'ONO]}{[RONO][R'OH]} \tag{6.42}$$

Alkyl nitrites (and also alkyl nitrates) have been much used as vasodilatory drugs, as indeed have other nitrosating agents such as the nitroprusside anion. This is discussed further in Chapter 7. There is strong evidence to suggest that the first stage in the vasodilatory action is the S-nitrosation of tissue-bound thiol groups (see equation (6.43)). The thionitrite produced is then believed to activate the enzyme guanylate cyclase which in turn brings about smooth muscle relaxation.

$$\sim\!\sim\!\sim SH + RONO \longrightarrow \sim\!\sim\!\sim SNO + ROH \tag{6.43}$$

6.5.3 *The Barton reaction and related reactions*

No discussion of the reactions of alkyl nitrites as nitrosating agents would be complete without reference to the reaction whereby a 1,5-rearrangement of the nitroso group in alkyl nitrites occurs from oxygen to carbon to give 4-nitroso alcohols.[68] The reaction is brought about photochemically and appears to involve (scheme (6.44)) homolysis of the alkyl nitrite to give the alkoxy radical which effects an intramolecular hydrogen atom abstraction to give the carbon radical. This then reacts with the nitric oxide to give the nitroso alcohol, which is usually found in its dimeric form or as the isomeric oxime. The reaction is thus an inter-molecular nitroso group rearrangement, which has been much used synthetically in the steroid field. The mechanism outlined in scheme (6.44) has been substantiated by a detailed analysis of the by-products, the formation of which can all be explained in terms of subsequent reactions of the proposed intermediate radicals.[69]

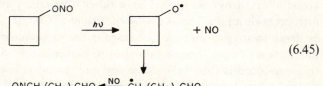

$$\tag{6.44}$$

Cyclic nitrites react similarly, both photochemically and thermally giving straight-chain aldehydes (see scheme (6.45)). Ring opening can occur from four-, five- and six-membered ring systems.[70] Under certain structural circumstances the proposed alkoxy radicals can break down (rather than rearrange) to give the more stable alkyl radicals which can react with nitric oxide to give nitrosoalkanes. Thus *t*-butyl nitrite forms nitrosomethane and acetone (see scheme (6.46)),[71] and one of the synthetic routes to nitrosocyclohexane uses the same principle.[72] It seems likely that a similar mechanism operates in the formation of nitrosotrifluoromethane, in the photolysis of trifluoroacetyl nitrite.[73]

$$\tag{6.45}$$

$$(CH_3)_3CONO \xrightarrow{h\nu} (CH_3)_3CO^{\cdot} + NO$$

$$(CH_3)_2CO + CH_3^{\cdot} \tag{6.46}$$

$$CH_3^{\cdot} + NO \longrightarrow CH_3NO$$

References

1. W.H. Hartung & F. Crossley, *Org. Synth.*, 1943, Coll. II, 363; W.L. Semon & V.R. Damerell, *ibid.*, 1943, Coll. II, 204; W.A. Noyes, *ibid.*, 1943, Coll. II, 108.
2. F.A. Lee & E.V. Lynn, *J. Am. Pharm. Soc.*, 1932, **21**, 125; H. Suginome & A. Osada, *J. Chem. Soc., Perkin Trans. 1*, 1982, 1963.
3. G. Olah, L. Noszko, I. Kuhn & M. Szelke, *Chem. Ber.*, 1956, **89**, 2374.
4. B. Brombeger & L. Phillips, *J. Chem. Soc.*, 1961, 5302.
5. S.A. Glover, A. Goosen, C.W. McCleland & F.R. Vogel, *S. Afr. J. Chem.*, 1981, **34**, 96.
6. A.D. Allen, *J. Chem. Soc.*, 1954, 1968.
7. D.H.R. Barton, J.M. Beaton, L.E. Geller & M.M. Pechet, *J. Am. Chem. Soc.*, 1961, **83**, 4076.

8. L.D. Hayward & R.N. Totty, *J. Chem. Soc., Chem. Commun.*, 1969, 997.

9. J. Casado, F.M. Lorenzo, M. Mosquera & M.F.R. Prieto, *Canad. J. Chem.*, 1984, **62**, 136.

10. S.E. Aldred, D.L.H. Williams & M. Garley, *J. Chem. Soc., Perkin Trans. 2*, 1982, 777.

11. J.H. Ridd, *Adv. Phys. Org. Chem.*, 1978, **16**, 1.

12. R.G. Pearson, *J. Chem. Educ.*, 1968, **45**, 581, 643.

13. M.R. Crampton, J.T. Thompson & D.L.H. Williams, *J. Chem. Soc., Perkin Trans. 2*, 1979, 18.

14. A.D. Yoffe & P. Gray, *J. Chem. Soc.*, 1951, 1412; M. Anbar & H. Taube, *J. Am. Chem. Soc.*, 1955, **77**, 2993.

15. E.H. White & W.R. Feldman, *J. Am. Chem. Soc.*, 1957, **79**, 5832.

16. R.G. Schweiger, *J. Org. Chem.*, 1976, **41**, 90.

17. R.G. Schweiger, *Fr. Demande*, 2, 313, 396, *Chem. Abstr.*, 1977, **87**, 103024.

18. A. Dalcq & A. Bruylants, *Tetrahedron Lett.*, 1975, **6**, 377.

19. V. Napoleone & Z.A. Schelley, *J. Phys. Chem.*, 1980, **84**, 17.

20. H. Niki, P.D. Maker, C.M. Savage & L.P. Breitenbach, *Int. J. Chem. Kinet.*, 1982, **14**, 1199; S. Koda, K. Yoshikawa, J. Okada & K. Akita, *Environ. Sci. Technol.*, 1985, **19**, 262.

21. C.A. Bunton, H. Dahn & L. Loewe, *Nature*, 1959, **183**, 163; H. Dahn, L. Loewe, E. Lüscher & R. Menasse, *Helv. Chim. Acta*, 1960, **43**, 287 and following papers.

22. S.S. Mirvish, L. Wallcave, M. Eagen & P. Shubik, *Science*, 1972, **177**, 65.

23. C.L. Walters, *Ind.-Univ. Co-op. Symp.*, 1974, 78.

24. D.L.H. Williams, *Food Cosmet. Toxicol.*, 1978, **16**, 365.

25. H. Marquardt, F. Rufino & J.H. Weisburger, *Science*, 1977, **196**, 1000.

26. P. Bogovski, *Environmental N-Nitroso Compounds*, Eds E.A. Walker, P. Bogovski and L. Gricuite, IARC Publications No. 14, IARC, Lyon, 1976, 3.

27. E.D. Hughes, C.K. Ingold & J.H. Ridd, *J. Chem. Soc.*, 1958, 88 and preceding papers.

28. A.B. Kyte, R. Jones-Parry & D. Whittaker, *J. Chem. Soc., Chem. Commun.*, 1982, 74.

29. R.E. Banks, R.N. Haszeldine & M.K. McCreath, *Proc. Chem. Soc.*, 1961, 64; C.W. Taylor, T.J. Brice & R.L. Wear, *J. Org. Chem.*, 1962, **27**, 1064.

30. H.A. Brown, N. Knoll & D.E. Rice, *U.S. Dept. Com., Office Tech. Serv.* AD 418, 1962, 638; *Chem. Abstr.*, 1964, **60**, 14709.

31. G. Stedman, *J. Chem. Soc.*, 1960, 1702.

32. J. Casado, A. Castro, M. Mosquera, M.F.R. Prieto & J.V. Tato, *Monatsh. Chem.*, 1984, **115**, 669.

33. K. Gleu & R. Hubold, *Z. Anorg. Chem.*, 1935, **223**, 305.

34. M. Kortum & B. Finckh, *Z. Phys. Chem.*, 1941, **B48**, 42.

35. M.N. Hughes & H.G. Nicklin, *J. Chem. Soc. A*, 1968, 450.

36. E. Halfpenny & P.L. Robinson, *J. Chem. Soc.*, 1952, 939.

37. E. Halfpenny & P.L. Robinson, *J. Chem. Soc.*, 1952, 928.

38. M. Anbar & H. Taube, *J. Am. Chem. Soc.*, 1954, **76**, 6243; *ibid.*, 1958, **80**, 1073; M. Anbar, *ibid.*, 1961, **83**, 2031.

39. D.J. Benton & P. Moore, *J. Chem. Soc. A*, 1970, 3179.

40. L.R. Dix & D.L.H. Williams, *J. Chem. Res. (S)*, 1982, 190.

41. P. Collings, K. Al-Mallah & G. Stedman, *J. Chem. Soc., Perkin Trans. 2,* 1975, 1734.
42. B.C. Challis & R.J. Higgins, *J. Chem. Soc., Perkin Trans. 2,* 1975, 1498.
43. W.G. Keith & R.E. Powell, *J. Chem. Soc. A,* 1969, 90.
44. D.H.R. Barton, R.H. Hesse, M.M. Pecket & L.C. Smith, *J. Chem. Soc., Chem Commun.,* 1977, 754.
45. W.A. Pryor, L. Castle & D.F. Church, *J. Am. Chem. Soc.,* 1985, **107**, 211.
46. W.M. Fischer, *Z. Phys. Chem.,* 1909, **65**, 61.
47. M.J.S. Dewar, M. Shanshal & S.D. Worley, *J. Am. Chem. Soc.,* 1969, **91**, 3590.
48. P.F. Dicks, S.A. Glover, A. Goosen & C.W. McCleland, *S. Afr. J. Chem.,* 1981, **34**, 101.
49. S. Oae, N. Asai & K. Fujimori, *J. Chem. Soc., Perkin Trans. 2,* 1981, 571.
50. M. Kobayashi, *Chem. Lett.,* 1972, 37.
51. E. Knoevenagel, *Chem. Ber.,* 1890, **23**, 2994.
52. *The Chemistry of Diazonium and Diazo groups,* Ed. S. Patai, Interscience, Chichester 1978; H. Zollinger, *Acc. Chem. Res.,* 1973, **6**, 335.
53. M. Stiles & R.G. Miller, *J. Am. Chem. Soc.,* 1960, **82**, 3802; L. Friedman & F.M. Logullo, *ibid.,* 1963, **85**, 1549; M. Stiles, R.G. Miller & U. Burckhardt, *ibid.,* 1963, **85**, 1792.
54. J.I.G. Cadogan, *J. Chem. Soc.,* 1962, 4257.
55. J.I.G. Cadogan, D.A. Roy & D.M. Smith, *J. Chem. Soc. C,* 1966, 1249.
56. A.J. Shenton & R.M. Johnson, *Int. J. Chem. Kin.,* 1972, **4**, 235.
57. M.J. Crookes & D.L.H. Williams to be published.
58. S.E. Aldred & D.L.H. Williams, *J. Chem. Soc., Perkin Trans 2,* 1981, 1021.
59. G. Bhattacharjee, *Indian J. Chem. Sec. B,* 1981, **20**, 526.
60. A.D. Allen & G.R. Schonbaum, *Canad. J. Chem.,* 1961, **39**, 947.
61. B.C. Challis & D.E.G. Shuker, *J. Chem. Soc., Chem. Commun.,* 1979, 315.
62. J.H. Bayless, F.D. Mendicino & L. Friedman, *J. Am. Chem. Soc.,* 1965, **87**, 5790.
63. S. Oae, N. Asai & K. Fujimori, *J. Chem. Soc., Perkin Trans. 2,* 1978, 1124.
64. A.G. Giumanini, M.F. Caboni & G.C. Galletti, *Gazz. Chim. Ital.,* 1981, **111**, 515.
65. J. Casado, A. Castro, F. Lorenzo & F. Meijide, *Monatsh. Chem.,* 1986, **117**, 335.
66. A.D. Allen & G.R. Schonbaum, *Canad. J. Chem.,* 1961, **39**, 940.
67. M.P. Doyle, J.W. Terpstra, R.A. Pickering & D.M. LePoire, *J. Am. Chem. Soc.,* 1983, **48**, 3379.
68. D.H.R. Barton, J.M. Beaton, L.E. Geller & M.M. Pechet, *J. Am. Chem. Soc.* 1960, **82**, 2640; P. Kabasakalian, E.R. Townley & M.D. Yudis, *J. Am. Chem. Soc.,* 1962, **84**, 2716, 2719; R.H. Hesse, *Adv. Free Radical Chem.,* 1969, **3**, 83.
69. D.H.R. Barton, R.H. Hesse, M.M Pechet & L.C. Smith, *J. Chem. Soc., Perkin Trans. 1,* 1979, 1159.
70. P. Gray & A. Williams, *Chem. Rev.,* 1959, **59**, 239; P. Kabasakalian & E.R. Townley, *J. Org. Chem.,* 1962, **27**, 2918.
71. G.R. McMillan, J.G. Calvert & S.S. Thomas, *J. Phys. Chem.,* 1964, **68**, 116.
72. Teijin Ltd., Neth. Appl. 6, 500, 271; *Chem. Abstr.,* 1966, **64**, 1981.
73. H.A. Brown, N. Knoll & D.E. Rice, *U.S. Dept. Com., Office Tech. Serv.,* 1962, A.D. 418, 638; *Chem. Abstr.,* 1964, **60**, 14709.

7

S-Nitrosation

It is well known that the sulphur atom in many sulphur-containing compounds is prone to electrophilic attack, eg in alkylation and halogenation; it is to be expected therefore that such compounds should also be subject to electrophilic nitrosation. Indeed, a comparison with *O*-nitrosation suggests that sulphur compounds should react broadly in a similar way, and should be more reactive, since a sulphur atom is more nucleophilic than a corresponding oxygen atom. However, much less is known about *S*-nitrosation than *O*-nitrosation and the same is true of the chemistry of *S*-nitroso compounds compared with that of *O*-nitroso species. This is due in part to the greater instability generally of *S*-nitroso compounds, the S—N bond being more susceptible to homolytic fission than is the O—N bond. Another factor is probably the greater reactivity of sulphur compounds in nitrosation, which usually requires special fast reaction techniques for the measurement of rate constants. With the advent of better experimental methods there has been a significant increase in the interest in *S*-nitrosation generally. A number of *S*-nitroso compounds have been isolated and characterised, whilst others have been detected in solution. Many have interesting synthetic applications and others are believed to play an important role in biologically important areas. A number of mechanistic studies have also been carried out. The topic has been reviewed.[1]

7.1 Nitrosation of thiols

The synthesis of thionitrites (sometimes called *S*-nitrosothiols) from thiols and the range of nitrosating agents (nitrosyl chloride, dinitrogen tetroxide, dinitrogen trioxide, alkyl nitrite, nitrous acid etc.) is probably the best-known example of *S*-nitrosation (see equation (7.1)), and is the exact

$$\text{RSH} + \text{XNO} \longrightarrow \text{RSNO} + \text{HX} \qquad (7.1)$$

sulphur counterpart of the formation of alkyl nitrites from alcohols. The reaction has been known for a long time, but many of the products are unstable. Thionitrites are generally yellow, red or even green. Their physical properties compare reasonably well with those of the more widely studied alkyl nitrites, given the changes expected for the smaller electronegativity of sulphur than of oxygen. Colour formation has been used as a test for thiols, and more recently a quantitative test for nitrosyl sulphuric acid has been established,[2] based on its reaction with thioglycolic acid and measurement of the yellow absorbance. Thionitrites, in common with many S-nitroso species have a broad absorption band in the uv–visible region with a maximum at ca 330 nm and extinction coefficients of ca 10^3. Early examples of isolated thionitrites are phenyl thionitrite and ethyl thionitrite prepared by Tasker and Jones.[3] Particularly stable thionitrites are those where there are bulky groups attached to the carbon atom bound to the sulphur atom. Examples include t-butyl thionitrite[4] (**7.1**), triphenylmethyl

$$(CH_3)_3CSNO \qquad (C_6H_5)_3CSNO \qquad HO_2C(NHCH_3CO)CHC(CH_3)_2SNO$$

7.1   **7.2**   **7.3**

thionitrite[5] (**7.2**), and the thionitrite (**7.3**) derived from N-acetyl penicillamine.[6] The latter is indefinitely stable in the solid phase and decomposes only slowly in solution. Its structure was confirmed by a crystal structure analysis. Other thionitrites, including that derived from cysteine,[7] are difficult to isolate in the pure form, and have to be stored at low temperatures.

In principle, any nitrosating agent can be used to prepare thionitrites, but experimentally the most suitable procedures are either, to react dinitrogen tetroxide with an equimolar amount of the thiol at ca $-10\,°C$ in an inert solvent such as chloroform,[8] or to use t-butyl nitrite as the reagent, also in chloroform[9] (equation (7.2)). Both procedures give rapid quantitative yields.

$$(CH_3)_3CONO + RSH \xrightarrow{\text{chloroform}} (CH_3)_3COH + RSNO \qquad (7.2)$$

There are some reports[10] of thionitrite formation from thiols and nitric oxide, which is not generally active as a nitrosating agent. Since even traces of oxygen will allow the formation of dinitrogen tetroxide (a well-known nitrosating species), it is likely that dinitrogen tetroxide is the reagent in many cases. However, in basic solution, and when oxygen is rigorously excluded, nitric oxide does react,[11] probably with the thiolate anion, eventually giving the disulphide as outlined in scheme (7.3). When pH $\leqslant 4$ the reaction is much inhibited, but if oxygen is admitted, then again the disulphide is formed, this time probably via a conventional electrophilic nitrosation by dinitrogen tetroxide, involving subsequent reaction of the intermediate thionitrite. The reaction via nitric oxide is quite different from conventional electrophilic nitrosation, but the mechanism is analogous to that proposed[12] to account for disulphide formation from a nitrosamine and a thiolate anion. In that case, trace quantities of the intermediate free radical were detected by esr spectroscopy.

$$RSH + B^- \rightleftharpoons RS^- + BH$$
$$RS^- + NO \longrightarrow RS\text{---}\dot{N}\text{---}O^- \rightleftharpoons RS\text{---}\dot{N}\text{---}OH$$
$$2RS\text{---}\dot{N}\text{---}OH \longrightarrow RSN(OH)N(OH)SR \longrightarrow RSSR + N_2 + N_2O$$

$$(7.3)$$

With reactants containing both the thiol group and the amino group (eg cysteine), reaction occurs preferentially at sulphur. Cysteine itself gives the familiar yellow colour very rapidly with nitrous acid. The *S*-nitroso derivative can be isolated with difficulty,[7] but more usually the isolated product is cystine, a result which can be rationalised in terms of subsequent reactions of the thionitrite. However, in one case[13] the isolated products of nitrosation of cysteine, its methyl ester and penicillamine were cyclic products such as the thiurancarboxylic acid **7.4** (from cysteine). The

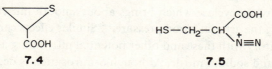

7.4   **7.5**

obvious interpretation is that ring closure occurs in the diazonium ion intermediate **7.5** by internal displacement of nitrogen by sulphur. This reaction pathway must involve *N*-nitrosation, which may arise under certain experimental conditions by rearrangement from the initially formed *S*-nitroso species. This is supported by the experimental observation of a strong green colour which first develops when penicillamine reacts with nitrous acid.[6] The cyclic product is recovered after a longer reaction time. Similar *S*- to *N*-rearrangements have been proposed to account for the products of nitrosation of thioureas, and also for the products of deamination of aliphatic amines containing the sulphide group. This is discussed further in sections 7.3 and 7.4.

It now seems likely that *S*-nitrosation plays an important part in the mechanism of action of a number of widely used vasodilatory drugs. These are used in the treatment of angina, heart failure and high blood pressure, and are also used to reduce blood pressure during surgical operations. Commonly used materials are alkyl nitrates (in particular glyceryl trinitrate), alkyl nitrites and the pentacyanonitrosylferrate (nitroprusside)

Table 7.1. *Values of k (equation (7.4)) for acid-catalysed nitrosation in water at 25 °C*

Substrate	$k/\text{l}^2\,\text{mol}^{-2}\,\text{s}^{-1}$	Reference
t-Butyl thiol	47[a]	15
Cysteine methyl ester	213	16
Cysteine	456, 443	17, 18
N-Acetylpenicillamine	840[b]	19(a)
Glutathione	1080	16
Mercaptosuccinic acid	1330	18
N-Acetyl cysteine	1590	16
Thioglycolic acid	2630	16
Mercaptopropanoic acid	4760	18

[a]In 50% dioxan–water
[b]At 31 °C

anion. It is now believed that each of these can effect *in vivo* S-nitrosation of tissue-bound thiol groups, and that the thionitrites thus formed activate the enzyme guanylate cyclase which brings about smooth muscle relaxation, thus reducing the arterial blood pressure.[14] Similar effects can be achieved in the laboratory with these and other potential nitrosating agents such as nitric oxide and sodium nitrite. Although S-nitrosation is deeply involved in this area, a detailed mechanistic picture awaits further work.

The kinetics of S-nitrosation of thiols follow a pattern familiar for a range of many other substrates. The rate equation (equation (7.4)) has been established for a range of thiols, by different research groups, and is interpreted as usual, in terms of rate-limiting attack by the nitrosonium ion or the nitrous acidium ion. Table 7.1 shows the values of the third-order rate constant k (from equation (7.4)) for a number of thiols. It is clear that all are very reactive; most of the data were obtained by stopped-flow spectrophotometry. The value of k for the more reactive thiols tends towards $7000\,\text{l}^2\,\text{mol}^{-2}\,\text{s}^{-1}$, which is taken to be the encounter-controlled limit for this type of reaction involving neutral substrates. Since thiols are not significantly protonated at moderate acidities, this means that the over-all reactivity of thiols is often greater than that of nucleophilic but basic amines where protonation reduces greatly the concentration of the reactive free amine form. There is scope for the use of thiols, particularly those containing the carboxylic acid group, as nitrous acid traps, for reactions in

$$\text{Rate} = k[\text{HNO}_2][\text{RSH}][\text{H}^+] \qquad (7.4)$$

water. Some have been tested,[18] and were found to be particularly effective, significantly more so than azide, which is regarded as one of the most reactive scavengers of nitrous acid.

In the case of cysteine, it has been established[19(b)] that two forms of the substrate are reactive *viz.* the zwitterion $HSCH_2CH(\overset{+}{N}H_3)COO^-$ and the *N*-protonated species $HSCH_2CH(\overset{+}{N}H_3)COOH$. Both react at rates close to the encounter limit with the zwitterion being the more reactive.

It is to be expected that thiols can compete effectively with amines in nitrosation reactions, and indeed it is possible to suppress completely nitrosamine formation from a secondary amine in the presence of a sufficient excess of either cysteine or *N*-acetylpenicillamine (see scheme (7.5)).[20]

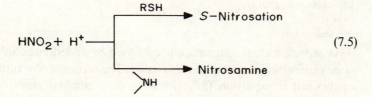

$$(7.5)$$

The indications are that *S*-nitrosation is faster than *O*-nitrosation, as expected on grounds of the different nucleophilicities. No exact comparisons have been made, but *N*-acetylpenicillamine **7.6** (a reasonably good

$$HSC(CH_3)_2CH(NHCH_3CO)COOH \qquad\qquad HOC(CH_3)_3$$

7.6 **7.7**

model for *t*-butyl thiol) is several orders of magnitude more reactive than is *t*-butanol **7.7**. The major difference between *O*- and *S*-nitrosation, however, is that the former is significantly reversible (see section 6.1) whilst the latter is effectively quantitative. The explanation for this is that oxygen is significantly more basic than sulphur generally in organic molecules ($\Delta pK_a \approx 5$), so that the rate of the reverse reaction (see scheme (7.6)) is greater for the oxygen case than for the sulphur case. The forward reaction (scheme (7.6)) is thus governed by the nucleophilicity ($S- > O-$), whereas the reverse reaction is governed by the basicity ($O- > S-$).

$$
\begin{aligned}
RSH + XNO &\rightleftharpoons RS\overset{+}{\underset{NO}{\overset{H}{\diagup}}} \rightleftharpoons RSNO + H^+ + X^- \\
ROH + XNO &\rightleftharpoons RO\overset{+}{\underset{NO}{\overset{H}{\diagup}}} \rightleftharpoons RONO + H^+ + X^-
\end{aligned}
\Bigg\} \quad (7.6)
$$

Table 7.2. *Values of* k_2 *for the reactions of nitrosyl chloride, nitrosyl bromide and nitrosyl thiocyanate with thiols in water at 25 °C*

Substrate	Nitrosyl chloride	k_2/l mol^{-1} s^{-1} Nitrosyl bromide	Nitrosyl thiocyanate
Cysteine methyl ester	1.1×10^6	4.6×10^4	7.5×10^2
Cysteine	1.2×10^6	5.5×10^4	6.5×10^2
Glutathione	1.2×10^7	5.5×10^5	3.9×10^3
Mercaptosuccinic acid	—	1.1×10^4	—
N-Acetyl cysteine	1.0×10^7	4.5×10^5	1.6×10^3
Thioglycolic acid	1.4×10^7	1.0×10^6	2.5×10^4
Mercaptopropanoic acid	—	4.5×10^5	—

As expected, thiol nitrosation is catalysed by added nucleophiles in the same general way as is amine nitrosation. The second-order rate constants k_2 (defined in equation (7.7)) for attack by nitrosyl chloride, nitrosyl bromide, and nitrosyl thiocyanate are given in table 7.2, for a number of thiols.[16,18] The trend of reactivity in the thiols parallels quite closely that

$$RSH + BrNO \xrightarrow{k_2} RS\overset{+}{\underset{NO}{\overset{H}{<}}} + Br^-$$

$$\text{Rate} = k_2[RSH][BrNO]$$

(7.7)

found for the uncatalysed reactions (table 7.1), and the familiar reactivity sequence nitrosyl chloride > nitrosyl bromide > nitrosyl thiocyanate is further demonstrated. Interestingly the *N*-acetyl derivative of cysteine is significantly more reactive than is cysteine itself, in all four nitrosation reactions. This suggests some degree of internal stabilisation of the positive charge on sulphur in the transition state, by the carbonyl oxygen atom, as shown in intermediate **7.8**, which can be taken as a reasonable model for the transition state.

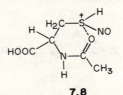

7.8

Another point of interest in table 7.2 is that the k_2 values for the most reactive reagent (nitrosyl chloride) and most reactive thiols do not approach the calculated encounter limit of $ca\ 7 \times 10^9\,\mathrm{l\,mol^{-1}\,s^{-1}}$, but appear to tend to a limit near $2 \times 10^7\,\mathrm{l\,mol^{-1}\,s^{-1}}$. This is in spite of the fact that for the uncatalysed reaction, it is believed that the reaction rate is close to the encounter limit. Secondary amines show the same pattern of behaviour with an apparent limit for k_2 at $\sim 1 \times 10^7\,\mathrm{l\,mol^{-1}\,s^{-1}}$, over a wide range of amine pK_a values. The reason for this is not clear, and more data are really needed to established the point more firmly for the thiols, but it may well be that the true limit is of the order of $1 \times 10^7\,\mathrm{l\,mol^{-1}\,s^{-1}}$, and that the reason why aromatic amines do tend towards the higher limit of $7 \times 10^9\,\mathrm{l\,mol^{-1}\,s^{-1}}$, has something to do with the presence of the aromatic π-electron system.

For most of the thiols listed in table 7.2, reaction is strictly first-order in [RSH] over the range studied. An exception is the reaction of thioglycolic acid in the presence of either bromide ion or thiocyanate ion, where plots of the observed first-order rate constants (with $[\mathrm{RSH}] \gg [\mathrm{HNO_2}]$) against [RSH] are initially linear, but curve downwards and tend to level off at higher [RSH], so that eventually reaction is zero-order in [RSH]. Under these circumstances the rate of formation of nitrosyl bromide (or nitrosyl thiocyanate) is rate-limiting, and is achieved (see scheme (7.8)) when $k_2[\mathrm{RSH}] \gg k_{-1}[\mathrm{H_2O}]$. Values of k_1 thus obtained agree well with the literature values, obtained by a similar procedure with other very reactive species such as azide ion and aniline derivatives.

$$\mathrm{H_3O^+ + HNO_2 \rightleftharpoons H_2NO_2^+ + H_2O}$$

$$(7.8)$$

$$\mathrm{H_2NO_2^+ + SCN^- \underset{k_{-1}}{\overset{k_1}{\rightleftharpoons} ONSCN + H_2O \xrightarrow[RSH]{k_2} RSNO + SCN^-}}$$

Nitrososulphonamides also react with thiols in reactions which probably involve *S*-nitrosation. *N*-Methyl-*N*-nitroso-*p*-toluenesulphonamide gives a quantitative yield of cystine when treated with cysteine.[21] A red colour is first formed, suggesting that the initial product is *S*-nitrosocysteine which then reacts further as outlined in scheme (7.9). In this case it is not known whether the substrate reacts directly with the thiol, or whether prior hydrolysis (a fairly rapid process) to give nitrous acid, first occurs. However,

$$
\left.
\begin{aligned}
\mathrm{RSO_2N(CH_3)NO + R'SH} &\longrightarrow \mathrm{RSO_2N(CH_3)H + R'SNO} \\
\mathrm{R'SNO} &\longrightarrow \mathrm{R'S^{\bullet} + NO^{\bullet}} \\
\mathrm{2R'S^{\bullet}} &\longrightarrow \mathrm{R'SSR'} \\
\mathrm{R'S^{\bullet} + R'SNO} &\longrightarrow \mathrm{R'SSR' + NO^{\bullet}} \\
\mathrm{R'SH + NO^{\bullet}} &\longrightarrow \mathrm{R'S^{\bullet} + HNO} \\
\mathrm{2HNO} &\longrightarrow \mathrm{N_2O + H_2O}
\end{aligned}
\right\} \quad (7.9)
$$

the denitrosation of nitrosamines is catalysed by both cysteine and glutathione, when reaction takes place in the presence of an excess of a nitrous acid trap,[22] so that here a direct *S*-nitrosation reaction occurs (see equation (7.10)). The reactivities of both cysteine and glutathione are comparable with that of chloride ion in this reaction. Similar reactions take place at other sulphur sites in thioureas and sulphides.

$$C_6H_5\overset{+}{N}H(CH_3)NO + RSH \longrightarrow C_6H_5NHCH_3 + R\overset{+}{S}\overset{NO}{\underset{H}{\diagup}} \quad (7.10)$$

The pentacyanonitrosylferrate ion also oxidises thiols to disulphides in alkaline solution[23] as outlined in equation (7.11). Again highly coloured (usually red) solutions are formed rapidly, and the colour then fades slowly. It is believed that an adduct is formed rapidly (see equation (7.12)), when the thiolate group is bound (via sulphur) to the nitrogen atom of the nitroso group. The decomposition of the adduct may take place by attack of thiolate anion, RS^-, or by some one-electron transfer process. The mechanism appears similar to that proposed for the nitrosation of amines by $Fe(CN)_5NO^{2-}$.

$$2[Fe(CN)_5NO]^{2-} + 2RS^- \longrightarrow RSSR + 2[Fe(CN)_5\overset{.}{N}O]^{3-} \quad (7.11)$$

$$[Fe(CN)_5NO]^{2-} + RS^- \rightleftharpoons [Fe(CN)_5NO(SR)]^{3-} \quad (7.12)$$

Both thio and dithiocarboxylic acids react with nitrous acid to form thionitrites,[24] by reaction with the thiol group as outlined in equations (7.13) and (7.14).

$$RC\overset{\diagup O}{\underset{SH}{\diagdown}} + HNO_2 \longrightarrow RC\overset{\diagup O}{\underset{SNO}{\diagdown}} + H_2O \quad (7.13)$$

$$RC\overset{\diagup S}{\underset{SH}{\diagdown}} + HNO_2 \longrightarrow RC\overset{\diagup S}{\underset{SNO}{\diagdown}} + H_2O \quad (7.14)$$

7.2 Thionitrites

The physical and chemical properties of thionitrites are well documented in a review by Oae and Shinhama.[25] In general, thionitrites decompose both thermally and photochemically to give the disulphide. Reaction with another thiol can give the unsymmetrical disulphide in excellent yield. Reduction gives the thiol, and oxidation the corresponding thionitrate. This section will concentrate only on the reactions of thionitrites which can be regarded as nitrosations.

Thionitrites will nitrosate secondary amines to give nitrosamines[26] (see equation (7.15)), and aniline derivatives[6] to yield azo dyes after coupling with β-naphthol.

$$RSNO + R_2'NH \longrightarrow R_2'NNO + RSSR \quad (7.15)$$

It is a matter of some concern that thionitrites formed *in vivo* under gastric conditions could in principle nitrosate secondary amines and amides in the lower intestine, producing carcinogenic nitrosamines and nitrosamides.

Thionitrites can be used to deaminate arylamines, and when the reaction is carried out in acetonitrile in the presence of anhydrous copper (II) halide, it is a good synthetic route to the corresponding aryl halides as outlined in equation (7.16).[25] Thionitrites can also bring about the nitrosation of alcohols forming alkyl nitrites, although the yields here are rather low.[27]

$$ArNH_2 + (CH_3)_3CSNO \xrightarrow[CH_3CN]{CuX_2} ArX + N_2 + (CH_3)_3CSSC(CH_3)_3$$
$$+ ((CH_3)_3C)_2S_3 \qquad (7.16)$$

None of these nitrosations brought about by thionitrites discussed so far has been studied mechanistically, so it is not known whether the nitrosation occurs directly or whether it involves some prior reaction of the thionitrite giving a free nitrosating species.

It is known that thionitrites can transfer the nitroso group directly to such nucleophiles as water (hydrolysis), halide ion, thiocyanate ion and thiourea, in acid-catalysed reactions (see for example equation (7.17)).[28] Reaction can only be brought about at relatively high acid concentration and only if the nitrosyl species (ONHal in equation (7.17)) is removed rapidly so that the reaction is irreversible. This behaviour is very similar to that found in the study of denitrosation reactions of nitrosamines (see chapter 5). Hydrolysis of thionitrites is much slower than that of alkyl nitrites, no doubt because of the greater basicity of the oxygen atom. The expected sequence, chloride ion < bromide ion < thiocyanate ion ≈ thiourea is observed experimentally.

$$RS\overset{+}{\underset{NO}{\overset{H}{<}}} + Hal^- \longrightarrow RSH + ONHal$$
$$\downarrow \qquad\qquad (7.17)$$
$$Removed$$

Mercuric ion also catalyses the hydrolysis of thionitrites; this reaction has been used as the basis of an analytical procedure for the quantitative estimation of thiols.[29] It has been suggested that a complex is formed

$$H_2O + \overset{O}{\underset{}{\underset{\diagdown}{\overset{\diagup}{N}}}}-\overset{+}{\underset{R}{\overset{Hg^+}{\underset{\diagdown}{S}}}} \longrightarrow HNO_2 + H^+ + RSHg^+$$
$$\downarrow \qquad\qquad (7.18)$$
$$Removed$$

between the thionitrite and the mercuric ion, which is then attacked by a water molecule, as outlined in equation (7.18).

7.3 Nitrosation of sulphides

The sulphur atom in a sulphide is expected to be at least as nucleophilic as that in a thiol, but since there is no convenient leaving group in the case of the sulphide, it is not surprising that thionitrites are not formed. There are, however, some indications that *S*-nitrosation of sulphides does occur. Disulphides react with dinitrogen tetroxide giving products whose formation can be rationalised by intermediate formation of both alkyl thionitrite and the sulphoxide ion, RSO^+, (equation (7.19)).[30] Photolysis of disulphides in the presence of nitric oxide similarly yields thionitrites (equation (7.20)).[31]

$$RSSR + N_2O_4 \longrightarrow RSNO + RSO^+$$

$$\downarrow \qquad\qquad \downarrow \qquad\qquad (7.19)$$

$$\text{Products} \qquad \text{Products}$$

$$CH_3SSCH_3 \xrightarrow{\ hv\ } 2CH_3S^{\cdot} \xrightarrow{\ NO\ } 2CH_3SNO \qquad (7.20)$$

Simple alkyl sulphides react with both nitrous acid and alkyl nitrites to give coloured solutions,[32] which suggests that some thionitrite species are formed, although no products have been isolated or characterised. Such species (which are capable of effecting further nitrosation of amines) have, however, been detected kinetically, by the observation of catalysis of nitrosation of *N*-methylaniline in acid solution by added dimethyl sulphide.[33] Catalysis is quite substantial and is comparable with that brought about by bromide ion. It seems likely that the *S*-nitroso ion $\gtrdot\overset{+}{S}\!-\!NO$, is formed in some equilibrium concentration, and then acts as a direct nitrosating agent (see equation (7.21)) in the same way as other XNO species are formed *in situ* and react further. Similarly, there is kinetic evidence that *S*-nitroso ions of this type can effect nitrosation *intramolecularly* in the presence of a suitably placed nucleophilic site such as the amino group. It has been found[34(a)] that the deamination reactions of both methionine and *S*-methylcysteine are very much faster than when the *S*-methyl group is not present in the molecule. The explanation (see scheme (7.22)) is that *S*-nitrosation occurs first, and is then followed by an internal rearrangement as outlined in scheme (7.22) for the reaction of *S*-methylcysteine. As noted earlier, *S*- to *N*-rearrangements of this type are quite well known in the chemistry of thioamides and thioureas. It follows that the rate of nitrosamine formation from a secondary amine which contains the thiolate group separated by two or three carbon atoms, should be greatly enhanced by the presence of the sulphide group. A similar explanation has been suggested to account for the greater reactivity of thioproline than of proline in nitrosation.[34(b)]

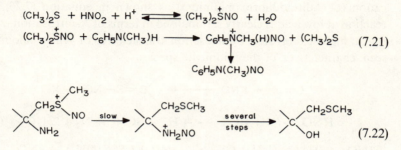

$$(CH_3)_2S + HNO_2 + H^+ \rightleftharpoons (CH_3)_2\overset{+}{S}NO + H_2O$$

$$(CH_3)_2\overset{+}{S}NO + C_6H_5N(CH_3)H \longrightarrow C_6H_5\overset{+}{N}CH_3(H)NO + (CH_3)_2S \qquad (7.21)$$

$$\downarrow$$

$$C_6H_5N(CH_3)NO$$

$$(7.22)$$

Molecular orbital calculations using the frontier orbital approach predict that reaction of nitrosonium ion at sulphur in an orbital-controlled reaction is favoured energetically over attack at nitrogen by a charge-controlled reaction.[35] This supports the general principle of HASB theory that the relatively soft nitrosyl cation (or its hydrated form) prefers reaction at the soft sulphur centre rather than at the harder nitrogen site.

Nitrosamines can also transfer the nitroso group to sulphides; methionine and S-methylcysteine are both reactive[22] and have about the same reactivity as the bromide ion (see scheme (7.23)).

$$C_6H_5N(CH_3)NO + H^+ \overset{fast}{\rightleftharpoons} C_6H_5\overset{+}{N}H(CH_3)NO$$

$$C_6H_5\overset{+}{N}H(CH_3)NO + RSR' \overset{slow}{\longrightarrow} C_6H_5NHCH_3 + R\overset{+}{S}\underset{R'}{\overset{NO}{<}}$$

$$R\overset{+}{S}\underset{R'}{\overset{NO}{<}} + NH_2NH_2 \overset{fast}{\longrightarrow} RSR' + \text{Decomposition products}$$

$$\left.\rule{0cm}{1.5cm}\right\} \quad (7.23)$$

7.4 Nitrosation of thiocarbonyl compounds

Thiocarbonyl compounds such as thioketones and thioureas are particularly reactive sulphur-nucleophiles. Thiourea for example is about as reactive as the iodide ion in S_N2 reactions at carbon, as measured by the Pearson nucleophilicity parameter.[36] In general thiocarbonyl compounds are nitrosated by XNO (equation (7.24)) to give initially the *S*-

nitrososulphonium ion. These *S*-nitroso ions are not very stable and react further, often ending up as the disulphide salts. By far the most widely studied reactions are those of thioureas. Early experiments by Werner[37] showed that two reactions were possible with thiourea itself, one at low acidities (equation (7.25)) leading to nitrogen and thiocyanate ion products, and the other, which dominates at higher acidities, giving the disulphide cation (*C,C*-dithiodiformamidinium) as shown in equation (7.26). The reaction at low acid is clearly rationalised in terms of *N*-nitrosation, whilst that at high acid involves *S*-nitrosation; transient red or yellow colours are seen, characteristic of the thionitrite species.

$$\mathrm{\diagdown\!\!\!\!\diagup} C{=}S + XNO \longrightarrow \mathrm{\diagdown\!\!\!\!\diagup} C{=}\overset{+}{S}{-}NO + X^- \qquad (7.24)$$

$$HNO_2 + (NH_2)_2CS \longrightarrow H^+ + SCN^- + N_2 + 2H_2O \qquad (7.25)$$

$$2HNO_2 + 2H^+ + 2(NH_2)_2CS \longrightarrow (NH_2)_2\overset{+}{C}SS\overset{+}{C}(NH_2)_2 + 2NO$$
$$+ 2H_2O \qquad (7.26)$$

It has been suggested[38] that even at low acidities reaction occurs at sulphur, followed by a *S*- to *N*-rearrangement. The evidence for this experimentally is kinetic, and depends upon the absence of a kinetic term in [H$^+$] in the rate equation for reaction leading to thiocyanate (equation (7.25)). Other experiments using ^{15}N-nmr measurements[39] have been interpreted in terms of a direct reaction at *N*, at low acidity, where it was possible to isolate the *N*-nitrosothiourea. The acid-catalysed hydrolysis of *N*-nitrosothioureas involves *N*- to *S*-migration.[40] However, the molecular orbital studies[35] referred to in section 7.3 again predict that nitrosation should occur preferentially at sulphur. Clearly two different reactions occur under the different experimental conditions and it has been further proposed that at high acidity a one-electron transfer mechanism occurs (equation (7.27)). This is very similar to the mechanism proposed for the reactions of cyclic sulphides with nitrosonium salts,[41] which is discussed in chapter 1.

$$(NH_2)_2CS + NO^+ \longrightarrow (NH_2)_2C{=}\overset{+\bullet}{S} + NO^\bullet$$
$$\Big\downarrow \qquad\qquad (7.27)$$
$$(NH_2)_2\overset{+}{C}SS\overset{+}{C}(NH_2)_2$$

Whatever the detailed mechanism of disulphide cation formation, it is another example of the oxidation of thiocarbonyl compounds to stable dications. A variety of chemical and electrochemical procedures[42] have been used for reactions of thioureas, thiocarbonates and thioketones (including heterocyclic examples). The structure of the dication has been established beyond doubt by crystal structure analyses of its salts.[43]

Under certain experimental conditions the final product of the nitrosation of thiourea is urea.[44] This occurs at the higher acidity and so probably involves a subsequent reaction of the S-nitroso thiouronium ion (see equation (7.28)), by nucleophilic attack of water, or by elimination of HSNO and subsequent hydration of the carbodiimide. This transformation can also be brought about by alkyl nitrites and a large range of other reagents including nitric acid, selenium dioxide etc., and is quite general for a range of thiocarbonyl and thiono compounds; this is probably one of the most convenient methods for the transformation of thiocarbonyl compounds to the corresponding carbonyl compounds.

$$\text{>C=S} + \text{XNO} \longrightarrow \text{>C=}\overset{+}{\text{S}}\text{—NO} \xrightarrow{\text{H}_2\text{O}} \text{>C=O} \quad (7.28)$$

The equilibrium constant for the formation of the yellow species derived from thiourea and nitrous acid (see equation (7.29)) has been determined as $5000 \, l^2 \, mol^{-2}$ at $25\,°C$.[38] So, in contrast with the nitrosyl halide formation, substantial quantities of the nitrous acid are converted into the S-nitroso compound, which is to be expected from the greater nucleophilicity of thiourea. The relatively large value of K for equation (7.29) accounts for the marked catalysis of nitrosation by thiourea discussed in chapter 1.

$$(\text{NH}_2)_2\text{CS} + \text{H}^+ + \text{HNO}_2 \rightleftharpoons (\text{NH}_2)_2\overset{+}{\text{C}}\text{SNO} + \text{H}_2\text{O} \quad (7.29)$$

The rate constant for the S-nitrosation of thiourea (k in equation (7.30)) is $69601 \, l^2 \, mol^{-2} \, s^{-1}$ at $25\,°C$,[17] and is taken to be that of the encounter-controlled reaction between the reagent (nitrosonium ion or nitrous acidium ion) and the thiourea molecule. Various N-methyl substituted thioureas also react at about the same rate (see table 7.3) and the overall activation energy of $65 \, kJ \, mol^{-1}$ is close to that expected for an encounter-controlled reaction.

$$\text{Rate} = k[(\text{NH}_2)_2\text{CS}][\text{HNO}_2][\text{H}^+] \quad (7.30)$$

Thioureas also react directly in acid solution with nitrosamines (equation (7.31))[45] forming initially the yellow-coloured S-nitroso species, which later form the disulphide salts. The reaction is a direct one, since it occurs in the presence of various nitrous acid traps. Alkyl nitrites[46] and thionitrites[28] behave similarly, as deduced from kinetic experiments on the denitrosation

Table 7.3. *Values of k (equation (7.30)) for acid-catalysed nitrosation in water at 25 °C, from reference 17*

Substrate	$k/l^2\,mol^{-2}\,s^{-1}$
Thiourea	6960
N-Methylthiourea	5620
N,N'-Dimethylthiourea	6610
N,N-Dimethylthiourea	5790
N,N,N'-Trimethylthiourea	4340

reactions using various nucleophiles. In all of these reactions thiourea is one of the most reactive of the nucleophiles studied.

$$R_2NNO + (NH_2)_2CS \xrightarrow{\ H^+\ } (NH_2)_2C\overset{+}{S}NO + R_2NH \qquad (7.31)$$

The subsequent reactions of the *S*-nitroso species derived from thiourea and its derivatives are discussed in chapter 1. The decomposition yielding the disulphide has been studied kinetically[47] and has been shown to involve two pathways, one a reaction between the *S*-nitroso species and thiourea, and another in which two molecules of the *S*-nitroso species react, eliminating nitric oxide and forming the dication.

7.5 Nitrosation of sulphinic acids

Sulphinic acids react with nitrous acid in the molar ratio of 2:1 to give sulphonyl hydroxylamine derivatives as outlined in equation (7.32).

$$2RSO_2H + HNO_2 \longrightarrow (RSO_2)_2NOH + H_2O \qquad (7.32)$$

Reaction is general for both alkyl and aryl sulphinic acids and occurs rapidly. Sulphinic acids are well-known *S*-nucleophiles,[48] reacting with a large range of electrophiles, at the sulphur atom rather than at the oxygen atom. It has been suggested[49] that the sulphinate ion, conventionally expressed by structure **7.9**, does in fact include a significant contribution

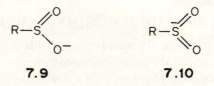

7.9   **7.10**

from **7.10** where the charge is located on sulphur.

The nitrosation reaction has been known for over a century[50] and has been used to characterise long-chain aliphatic sulphinic acids,[51] and also in the quantitative determination of sulphinic acids by direct titration with an acidic solution of sodium nitrite.[52] The reaction is also of some use in the synthesis of some hydroxylamine derivatives.[53] The stoichiometry of the reaction suggests, rather unusually, that two nitrosation steps are probably involved.

The kinetics of the reaction have been studied by stopped-flow spectrophotometry[54] (most S-nitrosations seem to occur very rapidly). There is good first-order dependence upon nitrous acid and also upon total stoichiometric concentration of the sulphinic acid, and the rates are the same for the disappearance of reactant as for appearance of product. Acid-catalysis, however, is not so straightforward. Above 0.06 M perchloric acid a plot of the rate constant (with $[RSO_2H] \gg [HNO_2]$) against $[H^+]$ is linear but with a substantial intercept. At lower acidities the plot curves towards the origin. This pattern is expected for a reaction which occurs by two parallel reactions, one involving the undissociated form and the other the anion form of the sulphinic acid as outlined in scheme (7.33). The expression for the first-order rate constant k_0 can readily be deduced from scheme (7.33) as that given in equation (7.34), where K_a is the acid dissociation constant of the sulphinic acid and $[Substrate]_T$ is its total stoichiometric concentration. In scheme (7.33) k_1 and k_2 are respectively the third-order rate constants (Rate $= k_1[H^+][HNO_2][RSO_2H] + k_2[H^+][HNO_2][RSO_2^-]$) for reaction with the undissociated form and the anion. Plots of $k_0(K_a + [H^+])/[H^+]$ vs. $[H^+]$ are linear and values of k_1 and k_2 readily determined as 820 and 11 800 $l^2 mol^{-2} s^{-1}$ respectively for benzenesulphinic acid. Thus the anion is the more reactive as expected, with a rate constant very similar to that of the nitrosation of thiocyanate ion. This suggests that both are at the encounter-controlled limit, which is expected to be somewhat larger than for reaction involving a neutral substrate.

$$RSO_2H \rightleftharpoons RSO_2^- + H^+$$

$$\downarrow HNO_2 \Big| k_1 \qquad\qquad \downarrow HNO_2 \Big| k_2 \qquad\qquad (7.33)$$

$$(RSO_2)_2NOH \qquad\qquad (RSO_2)_2NOH$$

$$k_0 = \frac{(k_1[H^+]^2 + k_2K_a[H^+])[Substrate]_T}{K_a + [H^+]} \qquad (7.34)$$

All the facts suggest that the mechanism is that set out in scheme (7.35) for the sulphinate anion. The first step is the rate-limiting S-nitrosation to give a nitrososulphinate, which rapidly effects another *S*-nitrosation of another sulphinate ion. Nitrososulphinate (or sulphonyl nitrite) intermediates have been postulated for some time, and have been isolated[55] from the reaction of sulphinic acids with dinitrogen tetroxide in ether at $-20\,°C$. They are very unstable compounds and are claimed to be the most reactive nitrosating species known (although there are no quantitative data available), reacting readily with amines,[56] alcohols[25] and thiols.[25]

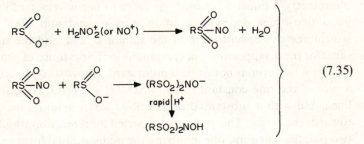

$$(7.35)$$

Like the nitrosation of amines and thiols, the nitrosation of sulphinic acids is catalysed by added non-basic nucleophilic species. The usual sequence of reactivity of the catalysts is followed, and again the sulphinate ion and the sulphinic acid molecule are reactive, the former being the more so, as expected.

Alkyl nitrites have also been shown[57] to yield sulphonyl hydroxylamine derivatives by reaction with sulphinic acids. The reaction is doubtless quite general for any nitrosating species.

7.6 Nitrosation of thiosulphate ion

It is long been known that acidic solutions of sodium nitrite give yellow solutions with thiosulphate ion. The product is clearly a *S*-nitroso species and has been described[58] as the nitrosothiosulphate anion (see equation (7.36)). It is rather unstable even in solution and decomposes to give nitric oxide and the tetrathionate ion (equation (7.37)).

$$HNO_2 + H^+ + S_2O_3^{2-} \longrightarrow O_3SSNO^- + H_2O \qquad (7.36)$$

$$2O_3SSNO^- = S_4O_6^{2-} + 2NO \qquad (7.37)$$

The rate law for nitrososulphinate ion formation (given in equation (7.38)[58]) is made up of two terms, the first representing rate-limiting nitrosation of the thiosulphate ion by nitrosonium ion or nitrous acidium ion and the second the rate-limiting formation of dinitrogen trioxide which then also effects nitrosation of the thiosulphate ion. The value of k_1 is $18\,000\,l^2\,mol^{-2}\,s^{-1}$, which is somewhat larger than the encounter-controlled limit of *ca* $7000\,l^2\,mol^{-2}\,s^{-1}$ for neutral substrates and *ca* $11\,000\,l^2\,mol^{-2}\,s^{-1}$ for singly-negatively-charged substrates. The trend is to be expected on electrostatic grounds, so it is likely that $18\,000\,l^2\,mol^{-2}\,s^{-1}$ represents the limit for such reactions involving doubly negatively charged substrates. The activation energy of $50\,kJ\,mol^{-1}$ also supports an encounter-controlled process.

$$\text{Rate} = k_1[\text{H}^+][\text{HNO}_2][\text{S}_2\text{O}_3^{2-}] + k_2[\text{HNO}_2]^2 \qquad (7.38)$$

The uv–visible spectrum of the yellow solution is very similar in shape and extinction coefficient to many other *S*-nitroso compounds. Analysis of the spectrum over a range of reactant concentrations in acetate buffers revealed a value of $1.66 \times 10^7\,l^2\,mol^{-2}$ for the equilibrium constant for nitrosothiosulphate ion formation at $25\,°C$. Such a large value (compared with those for such species as nitrosyl chloride, nitrosyl bromide and nitrosyl thiocyanate) suggests that if the ion has some reactivity as a nitrosating species itself, then catalysis of nitrosation of eg amines by thiosulphate should be possible. Such a prediction is borne out experiment-ally[59] for the nitrosation of *N*-methylaniline, hydrazine and sulphamic acid. Details are discussed in chapter 1.

7.7 Nitrosation of thiocyanate ion

Nitrosyl thiocyanate is readily formed from, in principle, any nitrosating agent and thiocyanate ion (see equations (7.39)). Synthetic procedures have included the reactions of nitrous acid with thiocyanic acid, nitrosyl chloride with silver thiocyanate, and ethyl nitrite with thiocyanic acid.[60] Nitrosyl thiocyanate is too unstable to be isolated and characterised fully in the pure state and is known only as a blood-red species, stable in solution at low concentration.[61] It is generally believed that the nitrogen atom of the nitroso group is bound to the sulphur atom of the thiocyanate group. Arguments based on HASB theory certainly favour bonding to sulphur and *ab initio* molecular orbital calculations[62] reveal that nitrosyl thiocyanate is significantly more stable than the isomer, nitrosyl isothiocyanate, where bonding is to nitrogen.

$$\left. \begin{array}{l} \text{XNO} + \text{SCN}^- = \text{ONSCN} + \text{X}^- \\ \text{H}^+ + \text{HNO}_2 + \text{SCN}^- = \text{ONSCN} + \text{H}_2\text{O} \end{array} \right\} \qquad (7.39)$$

The equilibrium constant for nitrosyl thiocyanate formation (equation (7.40)) has been determined spectrophotometrically;[63] at $25\,°C$ the value is $30\,l^2\,mol^{-2}$. The rate constant has not been measured directly but has been obtained, by the use of a very reactive substrate (see scheme (7.41)) at a sufficient concentration, so that the rate of nitrosyl thiocyanate formation is rate-limiting. Thus by using aniline and azide ion (both at low acid concentration), values of 1500 and 1460 l^2 mol^{-2} s^{-1} were obtained[64,65] for the third-order rate constant k (equation (7.42)) at 0 °C. These are very similar to values obtained for a range of other anions at the same temperature, which is readily explained in terms of encounter-controlled reactions. Later, values of 11 700 and 11 000 l^2 mol^{-2} s^{-1} were obtained at 25 °C using hydrazoic acid[66] and thioglycolic acid[16] respectively as the reactive substrates. Catalysis of nitrosation by thiocyanate ion with the intermediate involvement of nitrosyl thiocyanate is discussed in chapter 1.

$$K_{ONSCN} = \frac{[ONSCN]}{[HNO_2][SCN^-][H^+]} \tag{7.40}$$

$$\left.\begin{array}{l} HNO_2 + H^+ \rightleftharpoons H_2NO_2^+ \\ H_2NO_2^+ + SCN^- \rightleftharpoons ONSCN + H_2O \\ NOSCN + \text{Substrate} \longrightarrow \text{Nitrosation product} \end{array}\right\} \tag{7.41}$$

$$\text{Rate} = k[H^+][HNO_2][SCN^-] \tag{7.42}$$

As expected, nitrosamines,[67] alkyl nitrites[19] and thionitrites[38] also react in acid solution in their protonated forms with thiocyanate ion. This can easily be detected kinetically by a first-order dependence upon $[SCN^-]$ even in the presence of nitrous acid traps. This is shown in scheme (7.43) for the nitrosamine reaction, which will occur quantitatively to the denitrosation product in the presence of a nitrous acid trap, which removes the nitrosyl thiocyanate.

$$\left.\begin{array}{l} R_2N(CH_3)NO + H^+ \rightleftharpoons R_2\overset{+}{N}H(CH_3)NO \\ R_2\overset{+}{N}H(CH_3)NO + SCN^- \longrightarrow R_2NHCH_3 + ONSCN \\ \qquad\qquad\qquad\qquad\qquad\qquad\qquad \downarrow \\ \qquad\qquad\qquad\qquad\qquad\qquad\qquad \text{Removed} \end{array}\right\} \tag{7.43}$$

The effectiveness of thiocyanate as a nucleophile and of nitrosyl thiocyanate as a nitrosating agent is made use of in the *trans*-nitrosation reaction (equation (7.44)) of nitrosamines and amines[68] and also in the transfer reaction of the nitroso group from an alkyl nitrite to an amine[46] (see equation (7.45)).

$$R_2NNO + R_2'NH \xrightarrow[SCN^-]{H^+} R_2NH + R_2'NNO \tag{7.44}$$

$$RONO + R_2'NH \xrightarrow[SCN^-]{H^+} ROH + R_2'NNO \tag{7.45}$$

7.8 Nitrosation of bisulphite ion

Bisulphite ion reacts with nitrous acid to give hydroxylamine disulphonate (equation (7.46)). This reaction is important from a number of viewpoints; (a) it represents the first stage of the commercial Raschig synthesis of hydroxylamine,[69] (b) it is believed to be involved as one of the main reactions in the lead chamber process in the production of sulphuric acid,[70] and (c) it is important in atmospheric aerosol formation and in the chemical treatment of flue gases. The reaction itself has been known since 1894[71] and has been studied subsequently by a number of workers. Hydroxylamine is readily formed by acid hydrolysis of the disulphonate (equation (7.46)), probably by a stepwise mechanism.

$$H_2NO_2^+ \text{(or } NO^+) + HSO_3^- \longrightarrow HO_3SNO$$

$$HO_3SNO + HSO_3^- \longrightarrow [(HO_3S)_2NO]^- \xrightarrow{\ H^+\ } (HO_3S)_2NOH$$

$$(7.46)$$

The formation of hydroxylamine disulphonate clearly involves two *S*-nitrosation steps, and has a striking resemblance to the formation of hydroxylamine derivatives from sulphinic acids (section 7.5). A likely sequence (given in scheme (7.46)) involves the intermediate formation of nitrososulphonic acid, which, acting as a nitrosating agent, reacts with a further bisulphite ion to give the disulphonate. Kinetic studies are complicated by further sulphonation of the product to give a trisulphonate, hydrolysis of hydroxylamine disulphonate and also by hydrolysis of the proposed intermediate resulting in nitrous oxide formation. Nevertheless rate laws have been established in a number of independent studies.[72] These kinetic results support a mechanism outlined in scheme (7.47). In addition a kinetic term, second-order in bisulphate has been interpreted in terms of reaction via the metabisulphite ion $S_2O_5^{2-}$. A number of these studies also

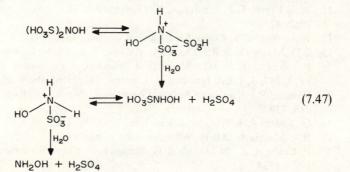

$$(7.47)$$

reported the presence of a bisulphite-independent kinetic term which has been attributed to rate-limiting nitrosonium ion formation. However, the most recent study failed to observe such a term in various buffer solutions, and has pointed out the dangers in interpretation of rather small positive intercepts in rate–concentration profiles. Most of these studies were carried out in the pH range 3–5.5 and it has not been possible to distinguish between nitrous acid and nitrite ion as possible nitrosating agents and bisulphite ion and sulphurous acid as the possible reactive forms of the substrate.

The nitrosation reactions of bisulphite and of sulphinic acids appear to represent the only clear-cut examples where the first-formed nitrosation product then reacts itself as a nitrosating agent, reacting with a further substrate ion or molecule, yielding in both cases hydroxylamine derivatives.

References

1. D.L.H. Williams, *Chem. Soc. Rev.*, 1985, **14**, 171.
2. M. Ziegler & O. Glemser, *Z. Analyt. Chem.*, 1955, **144**, 187; G. Röbisch & E. Ludwig, *Z. Chem.*, 1974, **14**, 103.
3. H.S. Tasker & H.O. Jones, *J. Chem. Soc.*, 1909, **95**, 1910, 1917.
4. G. Kresze & U. Uhlich, *Chem. Ber.*, 1959, **92**, 1048.
5. H. Rheinboldt, *Ber.*, 1926, **59**, 1311.
6. L. Field, R.V. Dilts, R. Ramanthan, P.G. Lenhert & G.E. Carnahan, *J. Chem. Soc., Chem. Commun.*, 1978, 249.
7. R. Bonnett & P. Nicolaidou, *J. Chem. Soc., Perkin Trans. 1*, 1979, 1969; M.J. Dennis, R.C. Massey & D.J. McWeeny, *J. Sci. Food Agric.*, 1980, **31**, 1195.
8. S. Oae, K. Shinhama, K. Fujimori & Y.H. Kim, *Bull. Chem. Soc. Jpn.*, 1980, **53**, 775.
9. M.P. Doyle, J.W. Terpstra, R.A. Pickering & D.M. LePoire, *J. Org. Chem.*, 1983, **48**, 3379.
10. L.J. Ignarro, B.K. Barry, D.Y. Gruetter, J.C. Edwards, E.H. Ohlstein, C.A. Gruetter & W.H. Baricos, *Biochem. Biophys. Res. Commun.*, 1980, **94**, 93.
11. W.A. Pryor, D.F. Church, C.K. Govindan & G. Crank, *J. Org. Chem.*, 1982, **47**, 156.
12. W.A. Waters, *J. Chem. Soc., Chem. Commun.*, 1978, 741.
13. C.D. Maycock & R.J. Stoodley, *J. Chem. Soc., Chem. Commun.*, 1976, 234.
14. P. Needleman, B. Jakschik & E.M. Johnson, *J. Pharmacol. Exp. Ther.*, 1973, **187**, 324; L.J. Ignarro, H. Lippton, J.C. Edwards, W.H. Baricos, A.L. Hyman, P.J. Kadowitz & C.A. Gruetter, *J. Pharmacol. Exp. Ther.*, 1981, **218**, 739.
15. G. Kresze & J. Winkler, *Chem. Ber.*, 1963, **96**, 1203.
16. P.A. Morris & D.L.H. Williams to be published.
17. P. Collings, K. Al-Mallah & G. Stedman, *J. Chem. Soc., Perkin Trans. 2*, 1975, 1734.

18. L.R. Dix & D.L.H. Williams, *J. Chem. Soc., Perkin Trans. 2*, 1984, 109.
19. (*a*) S.E. Aldred, D.L.H. Williams & M. Garley, *J. Chem. Soc., Perkin Trans. 2*, 1982, 777. (*b*) J. Casado, A. Castro, J.R. Leis, M. Mosquera & M.E. Pena, *J. Chem. Soc., Perkin Trans. 2*, 1985, 1859.
20. D.L.H. Williams & S.E. Aldred, *Food Chem. Tox.*, 1982, **20**, 79.
21. U. Schultz & D.R. McCalla, *Canad. J. Chem.*, 1969, **47**, 2021.
22. G. Hallett & D.L.H. Williams, *J. Chem. Soc., Perkin Trans. 2*, 1980, 624.
23. D. Mulvey & W.A. Waters, *J. Chem. Soc., Dalton Trans.*, 1975, 951; P.J. Morando, E.B. Borghi, L.M. de Schteingart & M.A. Blesa, *ibid.*, 1981, 435.
24. E.E. Reid, *Organic Chemistry of Bivalent Sulfur*, Chemical Publishing Co., New York, 1958, Vol. 1, p. 294.
25. S. Oae & K. Shinhama, *Org. Prep. Proced. Int.*, 1983, **15**, 165.
26. S. Oae, D. Fukushima & Y.H. Kim, *J. Chem. Soc., Chem. Commun.*, 1977, 407; M.J. Dennis, R. Davies & D.J. McWeeny, *J. Sci. Food Agric.*, 1979, **30**, 639; M.J. Dennis, R.C. Massey & D.J. McWeeny, *J. Sci. Food Agric.*, 1980, **31**, 1195.
27. S. Oae, Y.H. Kim, D. Fukushima & K. Shinhama, *J. Chem. Soc., Perkin Trans. 1*, 1978, 913.
28. S.S. Al-Kaabi, D.L.H. Williams, R. Bonnett & S.L. Ooi, *J. Chem. Soc., Perkin Trans. 2*, 1982, 227.
29. B. Saville, *Analyst.*, 1958, **83**, 670.
30. S. Oae, D. Fukushima & Y.H. Kim, *Chem. Lett.*, 1978, 279.
31. P.M. Rao, J.A. Copeck & A.R. Knight, *Canad. J. Chem.*, 1967, **45**, 1369.
32. E.M. Harper & A.K. Macbeth, *Proc. Chem. Soc.*, 1914, **30**, 15; A.K. Macbeth & D.D. Pratt, *J. Chem. Soc.*, 1921, **119**, 354.
33. T. Bryant & D.L.H. Williams, *J. Chem. Res. (S)*, 1987, 174.
34. (*a*) T.A. Meyer & D.L.H. Williams, *J. Chem. Soc., Chem. Commun.*, 1983, 1067. (*b*) T. Tahira, M. Tsuda, K. Wakabayashi, M. Nugao & T. Sugimura, *Gann.*, 1984, **75**, 889.
35. K.A. Jørgensen, *J. Org. Chem.*, 1985, **50**, 4758.
36. R.G. Pearson, H. Sobel & J. Songstad, *J. Am. Chem. Soc.*, 1968, **90**, 319.
37. A.E. Werner, *J. Chem. Soc.*, 1912, **101**, 2180; M.E. Coade & A.E. Werner, *J. Chem. Soc.*, 1913, **102**, 1221.
38. K. Al-Mallah, P. Collings & G. Stedman, *J. Chem. Soc., Dalton Trans.*, 1974, 2469.
39. J.W. Lown & S.M.S. Chauhan, *J. Org. Chem.*, 1983, **48**, 507, 513.
40. J.W. Lown & S.M.S. Chauhan, *J. Org. Chem.*, 1983, **48**, 3901.
41. W.K. Musker, T.L. Wolford & P.B. Roush, *J. Am. Chem. Soc.*, 1978, **100**, 6416.
42. R.L. Blankespoor, M.P. Doyle, D.M. Hedstrand, W.H. Tamblyn & D.A. Van Dyke, *J. Am. Chem. Soc.*, 1981, **103**, 7096.
43. O. Foss, J. Johnsen & O. Tvedten, *Acta Chem. Scand.*, 1958, **12**, 1782, and references therein.
44. K.A. Jørgensen, A.B.A.G. Ghattas & S.O. Lawesson, *Tetrahedron*, 1982, **38**, 1163; J.W. Lown & S.M.S. Chauhan, *J. Org. Chem.*, 1983, **48**, 507.
45. D.L.H. Williams, *J. Chem. Soc., Perkin Trans. 2*, 1977, 128.
46. S.E. Aldred & D.L.H. Williams, *J. Chem. Soc., Perkin Trans. 2*, 1981, 1021.

47. P. Collings, M. Garley & G. Stedman, *J. Chem. Soc., Dalton Trans.*, 1981, 331; M.S. Garley, G. Stedman & H. Miller, *J. Chem. Soc., Dalton Trans.*, 1984, 1959.
48. C.J.M. Stirling, *Int. J. Sulfur Chem. B*, 1971, **6**, 280; S. Oae & N. Kunieda, *Organic Chemistry of Sulfur*, Ed. S. Oae, Plenum, New York, 1977, p. 603.
49. E.N. Guryanova & Y.A. Syrkin, *Zh. Fiz. Khim.*, 1949, **23**, 105.
50. R. Otto & H. Ostrop, *Annalen*, 1867, **141**, 370; W. Koenigs, *Ber.*, 1878, **11**, 615.
51. C.S. Marvel & R.S. Johnson, *J. Org. Chem.*, 1948, **13**, 822.
52. B. Lindberg, *Acta Chem. Scand.*, 1963, **17**, 383 and references therein; J.P. Danehy & V.J. Elia, *Anal. Chem.*, 1972, **44**, 1281.
53. J.D. Birchall & C. Glidewell, *J. Chem. Soc., Dalton Trans.*, 1977, 10.
54. T. Bryant & D.L.H. Williams, *J. Chem. Soc., Perkin Trans. 2*, 1985, 1083.
55. S. Oae, K. Shinhama, K. Fujimori & Y.H. Kim, *Bull. Chem. Soc. Jpn.*, 1980, **53**, 775.
56. S. Oae, K. Shinhama & Y.H. Kim, *Bull. Chem. Soc. Jpn.*, 1980, **53**, 1065.
57. G. Kresze & W. Kort, *Chem. Ber.*, 1961, **94**, 2624.
58. M.S. Garley & G. Stedman, *J. Inorg. Nucl. Chem.*, 1981, **43**, 2863.
59. T. Bryant, D.L.H. Williams, M.H.H. Ali & G. Stedman, *J. Chem. Soc., Perkin Trans. 2*, 1986, 193.
60. E. Söderbäck, *Annalen*, 1919, **419**, 217; H. Lecher & F. Graf, *Ber.*, 1926, **59**, 2601.
61. C.C. Addison & J. Lewis, *Quart. Rev.*, 1955, **9**, 115.
62. K.A. Jørgensen & S.O. Lawesson, *J. Am. Chem. Soc.*, 1984, **106**, 4687.
63. G. Stedman & P.A.E. Whincup, *J. Chem. Soc.*, 1963, 5796.
64. C.A. Bunton, D.R. Llewellyn & G. Stedman, *J. Chem. Soc.*, 1959, 568.
65. G. Stedman, *J. Chem. Soc.*, 1959, 2949.
66. J. Fitzpatrick, T.A. Meyer, M.E. O'Neill & D.L.H. Williams, *J. Chem. Soc., Perkin Trans. 2*, 1984, 927.
67. I.D. Biggs & D.L.H. Williams, *J. Chem. Soc., Perkin Trans. 2*, 1975, 107.
68. S.S. Singer, *J. Org. Chem.*, 1978, **43**, 4612; S.S. Singer, G.M. Singer & B.B. Cole, *J. Org. Chem.*, 1980, **45**, 4931; S.S. Singer & B.B. Cole, *J. Org. Chem.*, 1981, **46**, 3461.
69. K.F. Purcell & J.C. Kotz, *Inorganic Chemistry*, W.B. Saunders Co., Philadelphia, 1977, p. 418; K. Jones in *Comprehensive Inorganic Chemistry*, Vol. 2, Eds., J.C. Bailar, H.J. Emeleus, R. Nyholm and A.F. Trotman-Dickenson, Pergamon Press, New York, 1973, p. 265.
70. F. Raschig, *Z. Angew. Chem.*, 1904, **17**, 1398.
71. E. Divers & T. Haga, *J. Chem. Soc.*, 1894, **65**, 523.
72. S.B. Oblath, S.S. Markowitz, T. Novakov & S.G. Chang, *J. Phys. Chem.*, 1982, **86**, 4852, and references therein.

8

Nitrosation involving metal complexes

Metal complexes or metal ions can, in some cases, allow nitrosation reactions to occur, which in the absence of such metal derivatives would not take place. This is achieved generally in one of two ways; (*a*) a new and more powerful nitrosating agent can be developed, such as the copper–nitrosyl complex in the catalysis by cupric ion of diethylamine nitrosation by nitric oxide,[1] or (*b*) by the oxidation of the substrate (such as an amine) to give a radical cation, which then can react with nitric oxide, as in the example of silver ion catalysis in amine nitrosation by nitric oxide.[2] Both of these examples are discussed in section 1.7, but not many other examples are known, although it has been reported that other ions such as ferrous, manganous and nickel ions can induce nitrosation of amines and alcohols, using again nitric oxide; no mechanistic details are known. This is an area of nitrosation chemistry which needs further development. Another example where a nitrosable substrate is generated concerns tertiary amines containing an α-hydrogen atom, which can be oxidised to the corresponding iminium ion (equation (8.1)) by mercuric ion.[3] These iminium ions (which can also be generated from amines and carbonyl compounds, see section 1.8) react readily with nitrite ion to form nitrosamines. Metal ions, notably molybdate ion, have also been shown[4] to be involved in the reduction of nitrate ion to nitrite ion enzymatically; the reduction can also be brought about electrochemically and also by stannous ion. Nitrite ion in acid solution can then of course effect the nitrosation of a range of substrates.

$$Hg^{2+} + R_2NCH_2R' = Hg + R_2\overset{+}{N}=CHR' + H^+$$
$$R_2\overset{+}{N}=CHR' + NO_2^- = R_2NNO + R'CHO \tag{8.1}$$

This chapter will concentrate on two aspects of the involvement of metal complexes in nitrosation; (*a*) nitrosation of nucleophiles coordinated to transition metals, and (*b*) the chemistry of metal nitrosyl complexes, in which they react as nitrosating agents.

8.1 Nitrosation of nucleophiles coordinated to metals

It appears that many nucleophilic species whch can be nitrosated in the free form, can also react when the nucleophilic species is present in a complex, coordinated to a metal. This is in spite of the fact that the nucleophilicity of such a group would be expected to be much reduced when coordinated to metals. Such groups include azide ion, ammonia, hydroxylamine, primary amines, secondary amines, and water. In some cases the nitrosation products are the same as when the substrate is not coordinated to metal, whereas in others, the nitrosation products remain coordinated to the metal.

8.1.1 Coordinated azide

One of the earliest reported reactions in this category refers to the nitrosation of the azide group present in a substitution-inert octahedral cobalt (III) complex. This reaction was studied by Haim and Taube[5] in 1963, and is shown in equation (8.2). The products here are the aquo complex, nitrogen and nitrous oxide. The gaseous products are the same as those obtained by the nitrosation of free azide. The mechanism is probably similar and involves the formation of the nitrosyl azide complex $(Co(NH_3)_5N_3NO^{3+})$ in the rate-limiting step. This complex then breaks down yielding nitrogen, nitrous oxide and the unstable pentacoordinated species $(Co(NH_3)_5^{3+})$ which reacts with water (or with any other nucleophile present) to give the observed product. The rate equation (equation (8.3)) is the same as that found for many substrates, and is interpreted as attack by nitrous acidium ion or nitrosonium ion. Catalysis by chloride ion, bromide ion and thiocyanate ion (and interestingly by nitrate ion and sulphate ion) indicates that the appropriate XNO species can also be reactive species. Nitrosation of coordinated azide appears to be general, and has also been noted for the following species $Rh(NH_3)_5N_3^{2+}$[6], $Co(en)_2(N_3)_2^+$[7], $Co(en)_2(N_3)H_2O^{2+}$[7], and $Cr(H_2O)_5N_3^{2+}$[8,9] (en is 1, 2-ethanediamine). In the case of the rhodium complex, an additional kinetic term appears, second-order in $[HNO_2]$, indicating attack by dinitrogen trioxide. Thus many of the features are present here, which also occur for the nitrosation of free azide ion or hydrazoic acid.

$$Co(NH_3)_5N_3^{2+} + HNO_2 + H^+ = Co(NH_3)_5OH_2^{3+} + N_2O + N_2$$

(8.2)

$$Rate = k[Co(NH_3)_5N_3^{2+}][HNO_2][H^+]$$

(8.3)

The collected rate data are in table 8.1. The differences in the rate constant values are relatively small, which suggests that reaction occurs

Table 8.1. *Values of k in Rate* $= k[HNO_2][H^+][Azide\ derivative]$ *in the reaction of nitrous acid with some azide derivatives.*

Reactant	$10^{-3}k/l^2\,mol^{-2}\,s^{-1}$	Temperature/°C	References
HN_3	0.034	0	10
N_3^-	2.5	0	11
$Co(NH_3)_5(N_3)^{2+}$	1.55	25	5
$Rh(NH_3)_5(N_3)^{2+}$	0.40	25	6
cis-$Co(en)_2(N_3)_2^+$	2.86	25	7
$trans$-$Co(en)_2(N_3)_2^+$	0.85	25	7
cis-$Co(en)_2(N_3)(H_2O)^{2+}$	0.44	25	7
$trans$-$Co(en)_2(N_3)(H_2O)^{2+}$	0.11	25	7
$Cr(H_2O)_5N_3^{2+}$	2.40, 2.49	25	8, 9

throughout near the encounter limit. This in turn suggests that in the coordinated species, reaction occurs at a terminal nitrogen, ie not the nitrogen atom coordinated to the metal, which would be expected to be very much less nucleophilic.

The structure of the intermediate is of some interest and has been discussed.[5] It has been shown, by ^{15}N labelling, ir and nmr spectroscopy,[12] that for the reaction of *trans*-$RuCl(das)_2N_3$ (where das is *o*-phenylene-bis-dimethylarsine) with nitrous acid, the intermediate nitrosyl azide derivative is cyclic in structure as outlined in structure **8.1**.

8.1

In some cases the final product is a dinitrogen (N_2) complex. For example, a ruthenium azide complex reacts with nitrosonium hexafluorophosphate[13] according to equation (8.4) giving a stable dinitrogen complex, whereas a ruthenium azide dinitrogen complex reacts with nitrous acid[14] to give a rather unstable bisdinitrogen species (equation (8.5)).

$$RuClN_3(das)_2 + NO^+ \longrightarrow RuClN_2(das)_2^+ \qquad (8.4)$$
$$Ru(en)_2N_3N_2^+ + HNO_2 \longrightarrow Ru(en)_2(N_2)_2^{2+} + N_2O \qquad (8.5)$$

8.1.2 Coordinated ammonia

The nitrosation of ammonia itself is a well-known reaction which is a convenient synthesis of nitrogen and is discussed in section 4.5. There are a number of examples in the literature where coordinated ammonia also undergoes nitrosation usually resulting in the formation of a dinitrogen complex. Scheidegger *et al*[15] in 1968 produced a bisdinitrogen complex of osmium from a monodinitrogen complex, (see equation (8.6)) where in effect one of the coordinated ammonia groups has been nitrosated. Later Buhr and Taube[16] described the nitrosation of $Os(NH_3)_4I_2^+$ which produces again a dinitrogen complex $Os(NH_3)_3I_2N_2^+$. Arguments were presented which suggested that the reactive nitrosating species is nitric oxide, formed when nitrous acid oxidises Os(III) to Os(IV) (see scheme (8.7)). A deprotonation reaction follows, and the nitric oxide reacts with this deprotonated form. Support for this mechanism comes from the observed products of the reaction of nitric oxide with Os(III) and Os(IV) complexes prepared independently.

Ruthenium compounds behave similarly. A dinitrogen complex is formed from $Ru(NH_3)_6^{3+}$ and nitric oxide in basic solution.[17] Kinetic measurements show the reaction to be first-order in $[Ru(NH_3)_6^{3+}]$, [NO] and $[OH^-]$, which suggests that a deprotonated form of one of the ammonia groups is a reactive intermediate. There is the analogy that sodamide ($NaNH_2$) reacts with nitrosyl chloride to give nitrogen.[18]

$$Os(NH_3)_5N_2^{2+} + HNO_2 \longrightarrow Os(NH_3)_4(N_2)_2^{2+} \qquad (8.6)$$

$$\left. \begin{array}{c} Os(NH_3)_4I_2^+ + HNO_2 + H^+ \rightleftharpoons Os(NH_3)_4I_2^{2+} + NO + H_2O \\ Os(NH_3)_4I_2^{2+} \rightleftharpoons Os(NH_3)_3I_2(NH_2)^+ + H^+ \\ Os(NH_3)_3I_2(NH_2)^+ + NO \longrightarrow Os(NH_3)_3I_2N_2^+ + H_2O \end{array} \right\} \quad (8.7)$$

8.1.3 Coordinated amines

Since both coordinated ammonia and azide can be nitrosated it is to be expected that amines generally will react similarly, and indeed, although the subject has not been studied systematically, there are a number of cases reported in the literature. All the evidence suggests that reaction occurs, as for ammonia, via the deprotonated form. It is well known that the acidity of amine protons is much increased by coordination to metal, and many deprotonated complexes have been isolated from a range of electrophiles.[19] Thus primary amines behave as if they were secondary amines. The nitrosation of 1, 2-ethanediamine in a platinum (IV) complex yields[20] a mono *N*-nitrosated species (equation (8.8)), identified as a ligand. The dianion of *N*, *N*′-dinitroso-1, 2-ethanediamine was obtained[21] by the dinitrosation of 1, 2-ethanediamine coordinated to platinum, this time in a bipyridyl complex, where the acidity of both protons in the amine is increased sufficiently to allow nitrosation to occur. Rather unusually the reaction of a ruthenium 1, 2-ethanediamine complex with nitric oxide in basic solution yields[22] the dinitrogen complex (equation (8.9)), where *N*-nitrosation and deamination of the primary amine has taken place.

$$Pt(en)_2Cl_2^{2+} \xrightarrow{\text{KNO}_2} Pt(en)[N(NO)CH_2CH_2NH_2]Cl_2^+ \qquad (8.8)$$

$$Ru(en)_3^{3+} + NO \xrightarrow{\text{OH}^-} Ru(en)_2N_2H_2O^{2+} \qquad (8.9)$$

Blackmer *et al.*,[23] were able to isolate the *N*-nitrosamine adduct from a methyleneaniline cobaloxime derivative (shown in outline in equation (8.10)). The free nitrosamine was liberated by treatment with chromous ion.

$$(Co)-CH_2-NHC_6H_5 + HNO_2 \longrightarrow (Co)-CH_2-N(NO)C_6H_5$$
$$\xrightarrow{\text{Cr}^{2+}} CH_3N(NO)C_6H_5 \qquad (8.10)$$

8.1.4 Coordinated hydroxylamine

Hughes *et al.*[24] found that hydroxylamine coordinated to cobalt reacted with nitrous acid to give the aquo complex and both nitrous oxide and nitrogen gases (equation (8.11)). Uncoordinated hydroxylamine gives only nitrous oxide as the gaseous product. The kinetics are more complicated in the case of the cobalt compounds, but above 0.1 M H^+ the familiar rate equation (equation (8.12)) was obeyed, indicating rate-limiting attack by nitrous acidium ion or nitrosonium ion. It is proposed that *O*-nitrosation probably first occurs giving a $Co(NH_2ONO)$ species which rearranges and breaks down by two separate pathways, one giving nitrous oxide and the other nitrogen. Support for this mechanism comes from the fact that the corresponding *O*-methylhydroxylamine derivative ($Co(en)_2Cl(NH_2OCH_3)^{2+}$) is unreactive under the same conditions.

$$Co(en)_2Cl(NH_2OH)^{2+} + HNO_2 \longrightarrow Co(en)_2ClH_2O^{2+} + N_2O + N_2 \qquad (8.11)$$

$$\text{Rate} = k[\text{Complex}][\text{HNO}_2][\text{H}^+] \qquad (8.12)$$

8.1.5 O-Nitrosation

One example[25] of *O*-nitrosation in a coordinated system occurs with aquopentamminecobalt (III). Reaction is believed to occur at the oxygen atom of a coordinated water molecule, to give the nitrito complex, which then rearranges to the nitro form (equation (8.13)). This sequence is confirmed by ^{18}O studies. The kinetics showed a second-order dependence upon nitrous acid concentration, and a first-order dependence upon the complex concentration, which is consistent with rate-limiting attack of dinitrogen trioxide.

$$Co(NH_3)_5H_2O^{3+} + N_2O_3 \longrightarrow Co(NH_3)_5ONO^{2+}$$

$$Co(NH_3)_5ONO^{2+} \longrightarrow Co(NH_3)_5NO_2^{2+} \qquad (8.13)$$

8.1.6 C-nitrosation

Nitrosation of a methylene group coordinated to cobalt has been reported,[26] in the case of nitrosation with nitrous acid of the pyridiomethyl cobalt (III) ion (see equation (8.14)). Here the pyridinium oxime derivative is formed as the free species, not coordinated to cobalt.

$$H\overset{+}{N}C_5H_4CH_2Co(CN)_5^{3-} \xrightarrow[H^+]{HNO_2} H\overset{+}{N}C_5H_4CH{=}NOH \qquad (8.14)$$

8.2 Metal nitrosyls as electrophilic nitrosating reagents

Metal nitrosyl complexes have been known for a long time but were not well characterised until the 1960s. Since that time there has been much work in the area and hundreds of such complexes have been prepared and their reactions studied. Much of the interest has centred around the bonding in nitrosyls and the comparison with corresponding carbonyls. In addition some nitrosyls have shown some interesting catalytic features in homogeneous systems, and one transition metal nitrosyl in particular, the pentacyanonitrosylferrate (II) (nitroprusside) ion $Fe(CN)_5NO^{2-}$, has considerable biological utility as a hypertensive reagent. Many useful and comprehensive reviews exist in this area; the chemistry of nitrosyls generally is excellently covered up to 1972 by reviews by Johnson and McCleverty,[27] and by Connelly.[28] More recent articles include those by McCleverty[29] and Bottomley.[30] Specific articles dealing with the nitroprusside ion[31] and ruthenium nitrosyls[32] have also appeared.

Transition metal nitrosyls can conveniently be thought of as sources of NO^+ or NO (ie they are either two-electron acceptors, or one-electron donors) and many of the reactions they undergo can be rationalised using this formalism. This section will only be concerned with metal nitrosyls reacting as electrophilic nitrosating agents, ie as carriers of the nitrosonium ion. A study of these reactions led Bottomley[33] to propose that complexes where the N—O stretching frequency is greater than 1886 cm^{-1} (or where the force constant $F(N—O)$ is greater than 13.8 mdyn Å^{-1}), will act as sources of nitrosonium ion, ie will react with nucleophiles such as hydroxide ion, alkoxide ion, ammonia, hydrazine, hydroxylamine, azide ion, amines etc. In practice this turns out to be a reasonably good generalisation.

8.2.1 Synthesis

There is a wide variety of methods available for the synthesis of metal nitrosyls. One inorganic text book[34] quotes thirteen different procedures which have been used. Below in equations (8.15)–(8.19) are set out some well-established procedures which can properly be regarded as nitrosations at metal centres. These include the use of the usual nitrosating reagents (nitrosyl salts, nitrosyl halides, alkyl nitrites, nitrous acid), as well as nitric oxide itself, which does not usually behave as a 'normal' nitrosating agent. Nitrosation at metal centres has not been the subject of a systematic mechanistic study, but there is no reason to doubt that many of the features outlined elsewhere for N-, C-, O- and S-nitrosations, also apply here.

$$CoCl_2L_2 + NO = CoCl_2L_2NO \quad \text{(for a variety of ligands L)} \quad (8.15)$$

$$PtCl_4^{2-} + ClNO = PtCl_5NO^{2-} \quad (8.16)$$

$$Ir(CO)ClL_2 + NO^+ = Ir(CO)ClL_2NO^+ \quad \text{(for a variety of ligands L)} \quad (8.17)$$

$$Fe(CO)_3(PPh_3)_2 + RONO + H^+$$
$$= Fe(CO)_2(NO)(PPh_3) + CO + ROH \quad (8.18)$$

$$NaFe(CO)_4H + 2NaNO_2 + 3CH_3CO_2H$$
$$= Fe(CO)_2(NO)_2 + 2CO + 2H_2O + 3CH_3CO_2Na \quad (8.19)$$

The remainder of this chapter is devoted to a discussion of the reactions of specific metal nitrosyls where they act as electrophilic nitrosating species. The examples chosen are not meant to be comprehensive, but rather illustrative of the range studied.

8.2.2 Nitroprusside (pentacyanonitrosylferrate II) $Fe(CN)_5NO^{2-}$

This is by far the best-known and most widely studied metal nitrosyl complex. The salts have been known since about 1850 and much of the qualitative chemistry of the species and related compounds, was established shortly after that time. Much is now known about the structure of the ion and its salts, from uv, ir, Mössbauer and crystallographic studies.[31]

Reactions occur readily with amines and with ammonia. Primary amines give the expected deamination products, ie alcohols and alkenes, whilst

secondary amines form nitrosamines – sometimes as free products, and at
other times bound to the metal in a complex. Even though the reactions
have been known for a long time,[35] the stoichiometry was not established
until 1971.[36] This is given in equation (8.20) for the reaction of secondary
amines. With hydroxide ion nitroprusside is converted to the nitro complex
$(Fe(CN)_5NO^{2-} \rightleftharpoons Fe(CN)_5NO_2^{4-})$. The equilibrium has been well
studied[37] and is clearly important in any reaction of nitroprusside carried
out under aqueous alkaline conditions. The reaction can be thought of as a
nitrosation reaction, analogous to the hydrolysis of the nitrosonium ion
$(NO^+ + 2OH^- \rightleftharpoons NO_2^- + H_2O)$. At pH 11 for example about 40% of
the nitroprusside is present as the nitrosyl form, so that nitrosation
reactions are generally carried out at pH < 10. With ammonia, the reaction
is essentially the same as that with hydroxide ion, but the nitro group is
replaced by ammonia. At high ammonia concentrations, however, nitrogen
gas is formed which strongly suggests that nitrosation of ammonia has
taken place.

$$Fe(CN)_5NO^{2-} + 2R_2NH = Fe(CN)_5NR_2H^{3-} + R_2NNO + H^+ \quad (8.20)$$

These reactions are best carried out in neutral or slightly basic media, as
distinct from many of the nitrosation reactions discussed in chapter 1 which
require acid conditions. This allows the ready synthesis of nitrosation
products using nitroprusside, if acid conditions need to be avoided.
Generally the complexes are highly coloured and reactions are often
accompanied by interesting colour changes. Such changes have been noted
for the reactions of nitroprusside with many other nitrogen containing
species which also contain a removable proton. These include indole,
pyrrole, phenylhydrazine and hydrazones. It is likely that nitrosation also
occurs here, but the products have not been characterised.

The nitrosation of primary and secondary amines by nitroprusside has
recently been examined kinetically by three independent groups.[38,39,40]
The observed rate equation is that given by equation (8.21), which includes
two terms, both first-order in nitroprusside concentration, and one first-
order in amine and the other second-order in amine. Sometimes, depending
on the experimental conditions and on the amine structure, either term can
dominate. In addition the reaction between nitroprusside and hydroxide
ion can result in another term in the rate equation, particularly at high pH.
This other reaction will not be discussed further. The variation of rate
constant with pH is interpreted in terms of the protonation equilibrium of
the amine, with reaction taking place only via the free-base form. The rate
data are readily interpreted in terms of an outline mechanism given in
scheme (8.22), in which a very low equilibrium concentration of an

intermediate is formed in which the amine group is bound to the nitrogen atom of the nitroso group. This intermediate then reacts competitively in rate-limiting steps with the amine or a water molecule to give the nitrosamine product and either the amine or aquo complex. However, it has been argued[39] that the attack on the amine occurs concurrently with a proton transfer to another amine molecule, ie that it is a concerted mechanism. Evidence in favour of the concerted process comes from a comparison of the second step in scheme (8.22) with the substitution of nitrite ion by a secondary amine in the corresponding nitro complex and also from the form of the detailed rate equation for a number of secondary amines with a range of pK_a values. A similar mechanism probably applies to primary amine reactions, where now the dialkylnitrosamine, R_2NNO, is replaced by the primary nitrosamine, $RNHNO$ which decomposes rapidly to give alcohol and nitrogen.

$$\text{Rate} = k_1[\text{Fe(CN)}_5\text{NO}^{2-}][\text{Amine}] + k_2[\text{Fe(CN)}_5\text{NO}^{2-}][\text{Amine}]^2$$

$$(8.21)$$

$$\text{Fe(CN)}_5\text{NO}^{2-} + \text{R}_2\text{NH} \underset{}{\overset{K}{\rightleftharpoons}} \left[\text{Fe(CN)}_5\text{N}\overset{\displaystyle O^-}{\underset{\displaystyle \overset{+}{N}HR_2}{}} \right]^{2-}$$

$$\text{Fe(CN)}_5\text{N}\overset{\displaystyle O^-}{\underset{\displaystyle \overset{+}{N}HR_2}{}} + \text{R}_2\text{NH} \xrightarrow{\text{slow}} \text{Fe(CN)}_5\text{NR}_2\text{H}^{3-} + \text{R}_2\text{NNO} + \text{H}^+$$

$$\text{Fe(CN)}_5\text{N}\overset{\displaystyle O^-}{\underset{\displaystyle \overset{+}{N}HR_2}{}} + \text{H}_2\text{O} \xrightarrow{\text{slow}} \text{Fe(CN)}_5\text{H}_2\text{O}^{3-} + \text{R}_2\text{NNO} + \text{H}^+$$

$$(8.22)$$

Nitrosation also occurs with mono and diaminoacids,[41] and as expected the reactivity of the amino group depends upon its basicity. For some diaminoacids, however,[42] an interesting cyclisation reaction occurs (which does not seem to occur when the reagent is nitrous acid) with nitroprusside; thus (see equation (8.23)) ornithine yields proline. No mechanistic details have been established.

$$\underset{\displaystyle NH_2}{CH_2}{-}CH_2{-}CH_2{-}\underset{\displaystyle NH_2}{CH}{-}COOH \xrightarrow{\text{Fe(CN)}_5\text{NO}^{2-}} \begin{array}{c} CH_2{-}CH_2 \\ | \quad\quad | \\ CH_2 \quad CH \\ \diagdown_{\displaystyle N}\diagup \quad COOH \\ | \\ H \end{array} \qquad (8.23)$$

Nitroprusside reacts with ketones and other compounds containing an acidic hydrogen atom. Usually a red colouration is rapidly formed which decomposes by a first-order process. The reaction involving acetone is the most widely studied[43] and it is believed to involve an intermediate complex formed from the carbanion derived from acetone (or possibly the enolate anion) and nitroprusside, as shown in scheme (8.24). The isolated products are the oxime and the aquo complex. More recently the kinetics of the formation of the coloured intermediate have been reported, for the reaction of nitroprusside with three carbanions.[44] For both ethylmalononitrile and diethyl malonate, the kinetics were consistent with a rate-limiting reaction between the carbanion and nitroprusside, but for malononitrile, reaction occurs between the doubly deprotonated anion and nitroprusside, as outlined in scheme (8.25).

$$(CH_3)_2CO + OH^- \rightleftharpoons \bar{C}H_2COCH_3 + H_2O$$

$$Fe(CN)_5NO^{2-} + \bar{C}H_2COCH_3 \longrightarrow \left[Fe(CN)_5N\overset{O}{\diagdown}_{CHCOCH_3}\right]^{4-} + H^+$$

$$\left[Fe(CN)_5N\overset{O}{\diagdown}_{CHCOCH_3}\right]^{4-} \xrightarrow{H_2O} [Fe(CN)_5H_2O]^{3-} + CH_3COCH\overset{}{\underset{NOH}{\|}}$$

$$(8.24)$$

$$^-CH(CN)_2 + OH^- \rightleftharpoons C(CN)_2^{2-} + H_2O$$

$$[Fe(CN)_5NO]^{2-} + [C(CN)_2]^{2-} \xrightarrow{slow} \left[Fe(CN)_5N\overset{O}{\diagdown}_{C(CN)_2}\right]^{4-}$$

$$(8.25)$$

There are a number of reactions of nitroprusside with sulphur-containing compounds, which are best rationalised in terms of *S*-nitrosation. These include the classical colour tests for the identification of both thiolate anion, HS^- and sulphite ion. Other sulphur species which form coloured solutions with nitroprusside include thiolate anion, RS^-, (derived from thiols), thiourea and alkyl thioureas. The formation of a purple-red solution from cysteine and nitroprusside in mildly alkaline solution is a well-known test for cysteine.

$$RSH + OH^- \rightleftharpoons RS^- + H_2O$$

$$[Fe(CN)_5NO]^{2-} + RS^- \rightleftharpoons \left[Fe(CN)_5N\overset{O}{\diagdown}_{SR}\right]^{3-}$$

$$2\left[Fe(CN)_5N\overset{O}{\diagdown}_{SR}\right]^{3-} \longrightarrow RSSR + 2[Fe(CN)_5\overset{\bullet}{N}O]^{3-}$$

$$(8.26)$$

$$
\left.
\begin{array}{l}
Fe(CN)_5NO^{2-} + SO_3^{2-} \rightleftharpoons \left[Fe(CN)_5N\diagdown^{O}_{SO_3}\right]^{4-} \\[20pt]
\left[Fe(CN)_5N\diagdown^{O}_{SO_3}\right]^{4-} \longrightarrow [Fe(CN)_5SO_3]^{5-} + ?
\end{array}
\right\}
\qquad (8.27)
$$

It is likely that many if not all of these reactions with sulphur-containing species proceed by the rapid and reversible formation of an adduct (which is the coloured species), where the sulphur compound is probably bonded to the nitrogen of the nitroso group. This cannot be stated with certainty, because the necessary structural analysis has not been carried out. These coloured intermediates usually decompose slowly, but the ultimate fate of the nitroso group is usually not known. Examples of the reaction with a thiol[45] and sulphite ion[46] are shown in schemes (8.26) and (8.27). The thiol is finally oxidised to the disulphide, but the structure of the iron complex product is not known, but may derive from the radical anion shown in scheme (8.26). In the reaction with sulphite ion, the final product derived from the nitroso group is not known, but may be nitrite ion. Again, the bonding in the intermediate complex is not known, but it has been shown[47] that the analogous reaction with a ruthenium nitrosyl does involve N—S bond formation.

Before leaving the nitroprusside ion, it is worth mentioning that it is a member of a group of materials which have been, and are being used medically to effect vasodepressor action. The other materials used are alkyl nitrites and alkyl nitrates. Specifically sodium nitroprusside solutions have been used since 1968, being injected into the blood stream during medical operations to reduce the level of the systemic arterial pressure. The mode of action of such vasodilators is not known even though their use has been widespread. However, evidence has accumulated[48] which suggests that they react with thiol groups to form *S*-nitroso compounds, which in turn markedly activate the enzyme guanylate cyclase; this then effects vascular smooth muscle relaxation. *S*-Nitrosothiols also bring about inhibition of human platelet aggregation.[49] Much more work is needed in this area before a full picture will emerge, but *S*-nitrosation by nitroprusside and other species is clearly important.

The use of nitroprusside has been restricted by the observation of symptoms of cyanide poisoning in patients, after treatment with nitroprusside, and there are reports of release of cyanide from nitroprusside in the presence of whole blood. One investigation[50] failed to find such release in the presence of whole blood, plasma and washed erythrocytes. This is to be expected from the high thermodynamic stability and a low-spin Fe(II) d^6 configuration, which should be very stable to ligand substitution. The reports of cyanide release can be explained in terms of a photochemical reaction which results in the formation of the aquo ion Fe(CN)$_5$H$_2$O^{2-}. This contains the low-spin Fe(III)d^5 configuration, which is subject to easy ligand substitution, and hence release of cyanide ion.

8.2.3 *Iron–sulphur cluster nitrosyls*

In 1858 the French chemist Roussin[51] prepared salts of the two ions Fe$_4$S$_3$(NO)$_7^-$ and Fe$_2$S$_2$(NO)$_4^{2-}$, now known as Roussin's black and red salts respectively. They have been of some interest from a structural point of view within the context of cluster chemistry. Their structures are now known from X-ray crystallographic data, and their spectroscopic properties are also of some interest. They undergo redox reactions and ligand substitution reactions and, in addition, compounds of this general type are now believed to be important biologically in cellular respiration and also probably in the antimicrobial action of nitrite. There is a review of the chemistry of these interesting compounds.[52]

Within the context of this chapter, these iron–sulphur clusters are also potential nitrosating agents since structural analysis reveals that the FeNO grouping is approximately linear, but the N—O vibrational frequency is less than 1850 cm^{-1}. However, it has been found[53,54] that the methyl ester of Roussin's red salt (**8.2**) will indeed nitrosate secondary amines such as

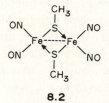

8.2

diethylamine, morpholine and pyrrolidine, to give nitrosamines, and it has been claimed that the reagent is more effective than is nitrite. Croisy *et al.*[54] found that nitrosation only occurred in the presence of air, which suggests that oxidation of the ester releases nitric oxide, or that nitric oxide once formed needs to be oxidised to the more reactive dinitrogen tetroxide or dinitrogen trioxide species. It may well be that this property accounts, at least partly, for the very high incidence of oesophageal cancer found for example in the Lixian province in China. In this area, vegetables are preserved in water and are covered with a mold, and have been shown to contain the methyl ester of Roussin's red salt. Coadministration of the ester and proline to rats resulted in the *in vivo* nitrosation of proline.[54]

8.2.4 *Ruthenium nitrosyls*

Apart from the nitroprusside ion, the ruthenium nitrosyls have been most studied in regard to nucleophilic attack at the nitrosyl group. Most ruthenium nitrosyls have $v(NO)$ greater than $1850 \, cm^{-1}$, which reflects the electron density at the metal; this is governed, among other factors by the nature and oxidation state of the metal. The RuNO group is almost linear in all complexes studied, and is set up for attack by nucleophiles at the nitroso nitrogen atom. There is a comprehensive account of the chemistry of ruthenium nitrosyls by Bottomley;[32] here we are concerned only with those reactions involving electrophilic nitrosation.

$$Ru(bipy)_2NOCl^{2+} + N_3^- = Ru(bipy)_2(H_2O)Cl^+ + N_2 + N_2O \tag{8.28}$$
$$(bipy = bipyridyl)$$

$$Ru(das)_2NOCl^{2+} + N_2H_4 = Ru(das)_2N_3Cl + 2H^+ + H_2O \tag{8.29}$$

$$Ru(NH_3)_5NO^{3+} + NH_2OH = Ru(NH_3)_5N_2O^{2+} + H^+ + H_2O \tag{8.30}$$

$$Ru(NH_3)_5NO^{3+} + 2RNH_2 = Ru(NH_3)_5N_2^{2+} + ROH + RNH_3^+ \tag{8.31}$$

$$Ru(bipy)_2NOCl^{2+} + ArNH_2 = Ru(bipy)_2N_2ArCl^{2+} + H_2O \tag{8.32}$$

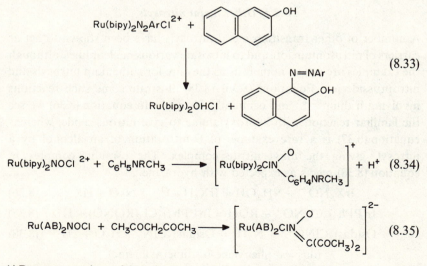

$$(8.33)$$

$$(8.34)$$

$$(8.35)$$

(AB = monoanion of 2, 4, 6-trioxo-5-oximinopyrimidine)

Equations (8.28)–(8.35) list examples of nitrosations of azide ion,[55] hydrazine,[56] hydroxylamine,[57] aliphatic amines,[58] aromatic amines[59] and β-diketones,[60] involving ruthenium nitrosyls. In most cases the product of nitrosation remains bound to the metal as in the reactions of hydrazine (equation (8.29)), hydroxylamine (equation (8.30)) and the amines (equations (8.31) (8.32) and (8.34)), whereas in the reaction with azide (equation (8.28)), nitrogen and nitrous oxide are released. The diazonium complex obtained from primary aromatic amines can react in a number of ways characteristic of diazonium ion, eg in the azo coupling reaction with β-naphthol (equation (8.33)). An interesting reaction occurs with secondary and tertiary aromatic amines where the 4-position in the aromatic ring becomes bonded to the nitroso nitrogen atom. This probably arises from initial attack at the amino nitrogen atom followed by a Fischer–Hepp rearrangement.[61] Equation (8.35) is an example of *C*-nitrosation, which occurs quite generally with a range of β-diketones and other compounds with acidic hydrogen atoms. Few of these reactions have been the subject of detailed mechanistic investigation, although the rate equations have been established in some cases. It is likely that the reaction mechanism is very similar to that for the nitroprusside reactions, involving the rapid formation of an equilibrium concentration of the adduct from the nitrosyl and the nucleophile.

8.2.5 Other metal nitrosyls

A number of other transition metal nitrosyls have been shown to act as carriers of nitrosonium ion, and to nitrosate various nucleophiles, although the examples are not so numerous as they are for ruthenium nitrosyls and nitroprusside ion. Equations (8.36)–(8.38) illustrate some such reactions involving iridium[62,63] and osmium[64] nitrosyls. In equation (8.36) we see the familiar reaction with hydroxylamine to give nitrous oxide, whereas equation (8.37) is a rare example of *O*-nitrosation of an alcohol by a nitrosyl, giving the alkyl nitrite complex. The osmium complex in equation (8.38) reacts as expected with hydrazine.

$$IrX_5NO^- + NH_2OH = IrX_5H_2O^{2-} + N_2O + H^+ \qquad (8.36)$$

$$Ir(PPh_3)_2Cl_3NO^+ + ROH = Ir(PPh_3)_2Cl_3(RONO) + H^+ \qquad (8.37)$$

$$Os(das)_2ClNO^{2+} + N_2H_4 = Os(das)_2ClN_3 + H_2O + 2H^+ \qquad (8.38)$$

(das = *o*-phenylene-bisdimethylarsine)

When nitric oxide is passed into anhydrous copper (II) halides in acetonitrile, soluble copper nitrosyl complexes are formed[65] (see equation (8.39)). These show characteristic $v(NO)$ frequencies, typical of linearly coordinated nitrosyls. Their reactions with aliphatic amines are complex, in that many products are formed, including unexpectedly geminal dihalides. Other products include alcohols and alkyl halides which probably are formed by nitrosation of the amine by the copper nitrosyl. Brackman and Smit[1] have also shown that nitrosation of amines and of alcohols can occur with nitric oxide in the presence of cupric ion. This reaction, which is discussed in section 1.7, is clearly very complex, but may be regarded overall as the catalytic breakdown of the dimer of nitric oxide, N_2O_2, into NO^- and NO^+. The NO^- fragment ends up as nitrous oxide and the nitrosonium ion is free to effect electrophilic nitrosation.

$$(CuX_2)_n + nNO \rightleftharpoons \tfrac{1}{2}n(CuX_2 \cdot NO)_2 \qquad (8.39)$$

It has been known for a long time,[66] that the pink colour given to cured meats by the addition of sodium nitrite, can be attributed to the formation of a nitric oxide–haemoprotein complex. These nitrosylhaems and nitrosylhaemo proteins have been much studied, from the structural point of view, and are examples of iron nitrosyl complexes. One of the commonest of such complexes is nitrosylprotohaem (see **8.3**) which has been syn-

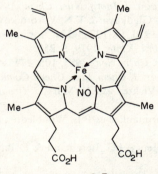

8.3

thesised and characterised as the dimethyl ester.[67] The basic structure is that of a porphyrin, with four pyrrole rings connected by methine bridges, with each nitrogen atom bound to the central iron atom.[68] Such nitrosylhaems are capable of effecting nitrosation of secondary amines under suitable conditions,[69] although it is not known whether this reaction occurs directly or indirectly by the intermediate formation of some other nitrosating species.

References

1. W. Brackman & P.J. Smit, *Recl. Trav. Chim. Pays-Bas*, 1965, **84**, 357, 372.
2. B.C. Challis & J.R. Outram, *J. Chem. Soc., Chem. Commun.*, 1978, 707.

3. N.L. Allinger, M.P. Cava, D.C. De Jongh, C.R. Johnson, N.A. Lebel & C.L. Stevens, *Organic Chemistry*, Worth, New York, 1971, p. 575.
4. E.I. Stiefel, *Proc. Nat. Acad. Sci.*, 1973, **70**, 988; G.P. Haight, *Acta Chem. Scand.*, 1961, **15**, 2012; *ibid.*, 1962, **16**, 221, 659.
5. A. Haim & H. Taube, *Inorg. Chem.*, 1963, **2**, 1199.
6. C.S. Davis & G.C. Lalor, *J. Chem. Soc. A*, 1968, 1095.
7. D. Loeliger & H. Taube, *Inorg. Chem.*, 1965, **4**, 1032.
8. R.C. Thompson & E.J. Kaufmann, *J. Am. Chem. Soc.*, 1970, **92**, 1540.
9. T.C. Matts & P. Moore, *J. Chem. Soc. A*, 1970, 2819.
10. G. Stedman, *J. Chem. Soc.*, 1959, 2943.
11. G. Stedman, *J. Chem. Soc.*, 1959, 2949.
12. P.G. Douglas & R.D. Feltham, *J. Am. Chem. Soc.*, 1972, **94**, 5254.
13. P.G. Douglas, R.D. Feltham & H.G. Metzger, *J. Am. Chem. Soc.*, 1971, **93**, 84.
14. L.A.P. Kane-Maquire, P.S. Sheridan, F. Basolo & R.G. Pearson, *J. Am. Chem. Soc.*, 1968, **90**, 5295.
15. H.A. Scheidegger, J.N. Armor & H. Taube, *J. Am. Chem. Soc.*, 1968, **90**, 3263.
16. J.D. Buhr & H. Taube, *Inorg. Chem.*, 1980, **19**, 2425.
17. S.D. Pell & J.N. Armor, *J. Am. Chem. Soc.*, 1973, **95**, 7625.
18. J.W. Mellor, *Inorganic and Theoretical Chemistry*, Longmans, Green and Co., London, 1928, Vol. 8, p. 257.
19. G.W. Watt & R.L. Hood, *J. Inorg. Nucl. Chem.*, 1970, **32**, 3359.
20. O.N. Adrianova, N.S. Gladkaya & V.N. Vorotnikova, *Russ. J. Inorg. Chem.* (English translation), 1970, **15**, 1278; O.N. Adrianova, A.S. Gladkaya and R.N. Shchelokov, *Koord. Khim.*, 1979, **5**, 255.
21. W.A. Freeman, *J. Am. Chem. Soc.*, 1983, **105**, 2725.
22. S.D. Pell & J.N. Armor, *J. Chem. Soc., Chem. Commun.*, 1974, 259.
23. G.L. Blackmer, T.M. Vickrey & J.N. Marx, *J. Organometallic Chem.*, 1974, **72**, 261.
24. M.N. Hughes, K. Schrimanker & P.E. Wimbledon, *J. Chem. Soc., Dalton Trans.*, 1978, 1634.
25. R.G. Pearson, P.M. Henry, J.G. Bergmann & F. Basolo, *J. Am. Chem. Soc.* 1954, **76**, 5920.
26. E.H. Bartlett & M.D. Johnson, *J. Chem. Soc. A*, 1970, 523.
27. B.F.G. Johnson & J.A. McCleverty, *Progr. in Inorg. Chem.*, 1966, **7**, 277.
28. N.G. Connelly, *Inorg. Chim. Acta Rev.*, 1972, **6**, 47.
29. J.A. McCleverty, *Chem. Rev.*, 1979, **79**, 53.
30. F. Bottomley, *Acc. Chem. Res.*, 1978, **11**, 158.
31. J.H. Swinehart, *Coord. Chem. Rev.*, 1967, **2**, 385.
32. F. Bottomley, *Coord. Chem. Rev.*, 1978, **26**, 7.
33. F. Bottomley, W.V.F. Brooks, S.G. Clarkson & S.B. Tong, *J. Chem. Soc., Chem. Commun.*, 1973, 919.
34. N.N. Greenwood & A. Earnshaw, *Chemistry of the Elements*, Pergamon, Oxford, 1984, pp. 515–6.
35. K.A. Hofmann, *Justus Liebigs Ann. Chem.*, 1900, **312**, 1; W. Manchot and P. Woringer, *Ber.*, 1913, **46**, 3514.

36. H. Maltz, M.A. Grant & M.C. Navaroli, *J. Org. Chem.*, 1971, **36**, 363.
37. J.H. Swinehart & P.A. Rock, *Inorg. Chem.*, 1966, **5**, 573.
38 L. Dozsa, A. Katho, V. Kormos & M.T. Beck, *Proc. Int. Conf. Coord. Chem.*, 19th, 1978, 35; L. Dozsa, V. Kormos and M.T. Beck, *Magy. Kem. Foly.*, 1983, **89**, 97.
39. J. Casado, M.A. Lopez Quintella, M. Mosquera, M.F.R. Prieto & J.V. Tato, *Ber. Busenges. Phys. Chem.*, 1983, **87**, 1208; J. Casado, M. Mosquera, M.F. Prieto & J.V. Tato, *ibid.*, 1985, **89**, 735.
40. A.R. Butler, C. Glidewell, J. Reglinski & A. Waddon, *J. Chem. Res. (S)*, 1984, 279.
41. A. Katho, Z. Bodi, L. Dozsa & M.T. Beck, *Magy. Kem. Foly.*, 1983, **89**, 102.
42. M.T. Beck, A. Katho & L. Dozsa, *Inorg. Chim. Acta*, 1981, **55**, L55.
43. J.H. Swinehart & W.G. Schmidt, *Inorg. Chem.*, 1967, **6**, 232; S.K. Wolfe & J.H. Swinehart, *Inorg. Chem.*, 1968, **7**, 1855.
44. A.R. Butler, C. Glidewell, V. Chaipanich & J. McGinnis, *J. Chem. Soc., Perkin Trans. 2*, 1986, 7.
45. D. Mulvey & W.A. Waters, *J. Chem. Soc., Dalton Trans.*, 1975, 951; P.J. Morando, E.B. Borghi, L.M. de Schteingart & M.A. Blesa, *ibid.*, 1981, 435.
46. A. Fogg, A.H. Norbury & W. Moser, *J. Inorg. Nucl. Chem.*, 1966, **28**, 2753.
47. F. Bottomley, W.V.F. Brooks, D.E. Paez & P.S. White, *J. Chem. Soc., Dalton Trans.*, 1983, 2465.
48. L.J. Ignarro, H. Lippton, J.C. Edwards, W.H. Baricos, A.L. Hyman, P.J. Kadowitz & C.A. Gruetter, *J. Pharmacol. Exp. Ther.*, 1981, **218**, 739.
49. B.T. Mellion, L.J. Ignarro, C.B. Myers, E.G. Ohlstein, B.A. Ballot, A.L. Hyman & P.J. Kadowitz, *Mol. Pharmacol.*, 1983, **23**, 653.
50. W.I.K. Bisset, A.R. Butler, C. Glidewell & J. Reglinski, *Br. J. Anaesth.*, 1981, **53**, 1015.
51. F.Z. Roussin, *Ann. Chim. Phys.*, 1858, **52**, 285; see also A.R. Butler, *J. Chem. Educ.*, 1982, **59**, 549.
52. A.R. Butler, C. Glidwell & M.H. Li, *Adv. Inorg. Chem.*, 1988, in the press.
53. G.H. Wang, W.X. Zhang & W.G. Chai, *Acta Chimica Sinica*, 1980, **38**, 95; M.Y. Wang, M.H. Li, Y.Z. Jiang, Y.H. Sun, G.Y. Li, W.X. Zhang, W.G. Chai & G.H. Wang, *Cancer Research on Prevention and Treatment*, 1983, **10**, 145.
54. A. Croisy, H. Ohshima & H. Bartsch, IARC *Sci. Publ.*, 1984, **57**, 327.
55. F.J. Miller & T.J. Meyer, *J. Am. Chem. Soc.*, 1971, **93**, 1294.
56. P.G. Douglas, R.D. Feltham & H.G. Metzger, *J. Am. Chem. Soc.*, 1971, **93**, 84.
57. F. Bottomley & J.R. Crawford, *J. Am. Chem. Soc.*, 1972, **94**, 9092.
58. C.P. Guengerich & K. Schug, *Inorg. Chem.*, 1978, **17**, 1378.
59. W.L. Bowden, W.F. Little & T.J. Meyer, *J. Am. Chem. Soc.*, 1973, **95**, 5084; *ibid.*, 1977, **99**, 4340.
60. C. Bremard, G. Nowogorocki & S. Sueur, *Inorg. Chem.*, 1979, **18**, 1549.
61. W.L. Bowden, W.F. Little & T.J. Meyer, *J. Am. Chem. Soc.*, 1976, **98**, 444.
62. F. Bottomley, S.G. Clarkson & S. Tong, *J. Chem. Soc., Dalton Trans.*, 1974, 2344.

63. C.A. Reed & W.R. Roper, *J. Chem. Soc., Dalton Trans.*, 1972, 1243.
64. F. Bottomley & E.M.R. Kiremire, *J. Chem. Soc., Dalton Trans.*, 1977, 1125.
65. M.P. Doyle, B. Siegfried & J.J. Hammond, *J. Am. Chem. Soc.*, 1976, **98**, 1627.
66. J. Haldane, *J. Hyg.*, 1901, **1**, 115.
67. J.C. Maxwell & W.S. Caughey, *Biochemistry*, 1976, **15**, 388.
68. R. Bonnett, S. Chandra, A.A. Charalambides, K.D. Sales & P.A. Scourides, *J. Chem. Soc., Perkin Trans. 1*, 1980, 1706, and earlier papers.
69. R. Bonnett, A.A. Charalambides, R.A. Martin, K.D. Sales & B.W. Fitzsimmons, *J. Chem. Soc., Chem. Commun.*, 1975, 884.

Index